WATER PRODUCTION FUNCTIONS FOR IRRIGATED AGRICULTURE

CENTER FOR
AGRICULTURAL
AND RURAL
DEVELOPMENT

Water Production Functions for Irrigated Agriculture

ROGER W. HEXEM, Agricultural Economist, Natural
Resource Economics Division, Economic Research
Service, United States Department of Agriculture.

EARL O. HEADY, C. F. Curtiss Distinguished Professor
in Agriculture; Professor of Economics; Director,
Center for Agricultural and Rural Development.
Iowa State University.

THE IOWA STATE UNIVERSITY PRESS / AMES, IOWA

© 1978 The Iowa State University Press
Ames, Iowa 50010. All rights reserved

Composed and printed in the U.S.A. by Science Press, Ephrata, Pa. 17522

First edition, 1978

Library of Congress Cataloging in Publication Data

Hexem, Roger W   1935–
    Water production functions for irrigated agriculture.

    Includes bibliographical references and index.
    1. Irrigated farming–Economic aspects–Mathematical
models. 2. Irrigated farming–Economic aspects–The
West–Mathematical models. 3. Production functions
(Economic theory) I. Heady, Earl Orel, 1916–
joint author. II. Title.
HD1714.H45   333.9'13'0978   77-17200
ISBN 0-8138-1785-4

C O N T E N T S

Preface, vii

List of Tables Appearing in Microfiche, ix

1. Role of Response Functions in Management and Decisions on Water Use, 3
2. Production Functions and Economic Applications, 9
3. Features of Selected Production Functions, 29
4. Procedures for Estimating Production Functions, 46
5. Origin and Features of the Research Project, 70
6. Analysis of Corn Experiments, 77
7. Analysis of Wheat Experiments, 106
8. Analysis of Cotton Experiments, 120
9. Single Product and Canonical Joint Product Functions for Sugar Beets, 136
10. Generalized Production Functions, 172
11. Derived Demand Functions for Water, 186
12. Product Supply Functions, 194
13. Programmed Water Demands, 200

Notes, 208

Index, 213

Microfiche: See inside back cover

v

This book deals with the estimation and application of water production functions. It is based on a fairly large number of controlled experiments in western states. The study was made possible through a grant by the Bureau of Reclamation to the Center for Agricultural and Rural Development of Iowa State University. The experiments were conducted cooperatively with a group of agronomists in Colorado, Texas, Kansas, Arizona, California, and Oregon (although data from Oregon are not presented). The study was possible only because of the interest of personnel at the Bureau of Reclamation and those persons and agricultural experiment stations listed in Chapter 5. Experiments were conducted for corn, cotton, wheat, sugar beets, and corn silage over the years 1968–71 inclusive.

The study was initiated because of the Bureau of Reclamation's interest in more efficient use of water for irrigation and the need to generalize knowledge of water productivity in terms of crop yields that could be related to different soils and regions in the states relevant to the Bureau of Reclamation's programs. The generalized production functions were sought so that experiments might not have to be conducted on every specific soil type for which irrigation water might be made available as farm plans were developed to estimate (a) the best farm organization and water use, (b) the yield increases expected, and (c) the relevant charges or repayment schedule to associate with each farm receiving water from an irrigation project.

Of course, water production functions are not readily and practically estimated apart from other management and environmental variables for these purposes. While water interacts with numerous input variables under the farmer's control (for example, rates of seeding, fertilization, pest control, etc.), yield response to water is estimated only in interaction with fertilizer applications in this study. This restriction was necessary because of the limited budget and time available for the study. For purposes of generalization of water production functions, it is necessary to quantify environmental variables representing climate and soils and incorporate them into the estimated production functions. Hence, these steps were attempted throughout the period of the study. Since the experiments had to be undertaken at locations of interested cooperating agronomists and the sites of their experimental facilities over a four-year period, a restricted sample of soil and climatic observations was available. With unlimited funds and time, an experiment of this nature would have "a stratified sample" of locations with a wide range of soil and climatic variables over which the experiments would be conducted. An experiment of this nature and expanse is needed but it should extend over a period of 20 years with a budget encompassing millions of dollars. The time and funds for the present study were small fractions of these amounts.

Through the excellent cooperation of the agronomists and institutions listed in Chapter 5, we believe that important progress was made in the objectives of this study. The results reported are encouraging and should be carried forward over more time and space by other research workers. The results should be encouraging especially for those large-scale studies of water and irrigation being financed by the Agency for International Development in less-developed countries. These projects have large funds and span many countries. If designed appropriately, they could give rise to water and fertilizer response functions that allow transfer of knowledge and generalization of yields to other countries and soil and climatic regimes.

There is growing need for improved knowledge of water response functions. A major reason is that very few water production functions have been estimated over the entire world—even though these functions are basic in the optimal allocation of water among soil types, farms, and uses. An immediate approach to optimal water development and allocation is needed if sufficient progress is to be made in meeting the world's problems of hunger and water quality as more chemicals and specialized cropping are used to extend food output. This book illustrates the need for and use of such data.

Since this book will be used by agronomists who are not all fully acquainted with the economic principles of water allocation and production function estimation, the first four chapters are devoted to explanation of the concepts and principles.

We express our appreciation to the Bureau of Reclamation for making this study possible. We are indebted to Ira Watson, Robert Struthers, John Maletic, W. W. Reedy, Robert Berger, and Allen Kleinman for the leadership and advice they provided to us.

Roger W. Hexem
Earl O. Heady

TABLES APPEARING IN MICROFICHE

NUMBER   TITLE

6.1. Summary of treatment combinations, plant population, and corresponding corn (grain) yields, in pounds per acre, at 15.5 percent moisture (Colby, Kansas, 1971)

6.2. Predicted corn (grain) yields, in pounds per acre, for designated water and nitrogen levels (Colby, Kansas, 1971)

6.3. Estimated marginal rates of substitution (MRS) between acre-inches of water and acres of land and between acre-inches of water and pounds of nitrogen at specified yield levels for corn (grain) (Colby, Kansas, 1971)

6.4. Summary of treatment levels and corresponding corn (grain) yields, in pounds per acre, at 15.5 percent moisture (Colby, Kansas, 1970)

6.5. Predicted corn (grain) yields, in pounds per acre, at 15.5 percent moisture, for designated water and nitrogen levels (Colby, Kansas, 1970)

6.6. Summary of treatment levels and corresponding corn (grain) and silage yields, in pounds per acre, at 15.5 and 75 percent moisture, respectively (Fort Collins, Colorado, 1968)

6.7. Predicted corn (grain) yields, in pounds per acre, at 15.5 percent moisture, for designated water and nitrogen levels (Fort Collins, Colorado, 1968)

6.8. Predicted corn (silage) yields, in pounds per acre, at 75 percent moisture, for designated water and nitrogen levels (Fort Collins, Colorado, 1968)

6.9. Summary of treatment levels, plant population, and corresponding corn (grain) yields, in pounds per acre, at 15.5 percent moisture (Davis, California, 1970)

6.10. Predicted corn (grain) yields, in pounds per acre, at 15.5 percent moisture, for designated water and nitrogen levels (Davis, California, 1970)

6.11. Summary of treatment levels, plant population, and corresponding corn (grain) yields, in pounds per acre, at 15.5 percent moisture (Davis, California, 1969)

6.12. Predicted corn (grain) yields, in pounds per acre, at 15.5 percent moisture, for designated water and nitrogen levels (Davis, California, 1969)

6.13. Summary of treatment levels and corresponding corn (grain) and silage yields, in pounds per acre, at 15.5 and 70.0 percent moisture, respectively (Mesa, Arizona, 1971)

6.14. Predicted corn (grain) yields, in pounds per acre, at 15.5 percent moisture, for designated water and nitrogen levels (Mesa, Arizona, 1971)

6.15. Predicted corn (silage) yields, in pounds per acre, at 70 percent moisture, for designated water and nitrogen levels (Mesa, Arizona, 1971)

6.16. Summary of treatment levels and corresponding corn (grain) and silage yields, in pounds per acre, at 15.5 and 70.0 percent moisture, respectively (Mesa, Arizona, 1970)

6.17. Predicted corn (grain) yields, in pounds per acre, at 15.5 percent moisture, for designated water and nitrogen levels (Mesa, Arizona, 1970)

6.18. Predicted corn (silage) yields, in pounds per acre, at 70 percent moisture, for designated water and nitrogen levels (Mesa, Arizona, 1970)

6.19. Summary of treatment levels and corresponding corn (grain) and silage yields, in pounds per acre (Yuma Mesa, Arizona, 1970)

6.20. Predicted corn (grain) yields, in pounds per acre, at 56 pounds per bushel, for designated water and nitrogen levels (Yuma Mesa, Arizona, 1970)

6.21. Predicted corn (silage) yields, in pounds per acre, for designated water and nitrogen levels (Yuma Mesa, Arizona, 1970)

6.22. Summary of treatment levels and corresponding corn (grain) and silage yields, in pounds per acre (Yuma Valley, Arizona, 1970)

6.23. Predicted corn (grain) yields, in pounds per acre, at 56 pounds per bushel, for designated water and nitrogen levels (Yuma Valley, Arizona, 1970)

6.24. Predicted corn (silage) yields, in pounds per acre, for designated water and nitrogen levels (Yuma Valley, Arizona, 1970)

6.25. Summary of treatment levels and corresponding corn (grain) yields, in pounds per acre, at 15.5 percent moisture (Safford, Arizona, 1971)

6.26. Predicted corn (grain) yields, in pounds per acre, at 15.5 percent moisture, for designated water and nitrogen levels (Safford, Arizona, 1972)

6.27. Summary of treatment levels, plant population, and corresponding corn (grain) yields, in pounds per acre, at 15.5 percent moisture (Plainview, Texas, 1971)

6.28. Irrigation dates and quantities of water applied in acre-inches for five irrigations in production of corn (grain) (Plainview, Texas, 1971)

6.29. Predicted corn (grain) yields, in pounds per acre, at 15.5 percent moisture, for designated water and nitrogen levels (Plainview, Texas, 1971)

6.30. Summary of treatment levels, plant population, and corresponding corn (grain) yields, in pounds per acre, at 15.5 percent moisture (Plainview, Texas, 1970)

6.31. Irrigation dates and quantities of water applied in acre-inches for five irrigations in production of corn (grain) (Plainview, Texas, 1970)

6.32. Predicted corn (grain) yields, in pounds per acre, at 15.5 percent moisture, for designated water and nitrogen levels (Plainview, Texas, 1970)

6.33. Summary of treatment levels and corresponding corn (grain) yields, in pounds per acre, at 15.5 percent moisture (Plainview, Texas, 1969)

6.34. Irrigation dates and quantities of water applied in acre-inches for five irrigations in production of corn (grain) (Plainview, Texas, 1969)

6.35. Predicted corn (grain) yields, in pounds per acre, at 15.5 percent moisture, for designated water and nitrogen levels (Plainview, Texas, 1969)

6.36. Summary of treatment levels, average plant population, and corresponding corn (grain) yields, in pounds per acre, at 15.5 percent moisture (Plainview, Lake site, Texas, 1971)

6.37. Irrigation dates and quantities of water applied in acre-inches for five irrigations in production of corn (grain) (Plainview, Lake site, Texas, 1971)

6.38. Predicted corn (grain) yields, in pounds per acre, at 15.5 percent moisture, for designated water and nitrogen levels (Plainview, Lake site, Texas, 1971)

7.1. Summary of treatment levels and corresponding wheat yields (Yuma Valley, Arizona, 1971–72)

7.2. Predicted wheat yields, in pounds per acre, for designated water and nitrogen levels (Yuma Valley, Arizona, 1971–72)

7.3. Summary of treatment levels and corresponding wheat yields (Yuma Valley, Arizona, 1970–71)

7.4. Predicted wheat yields, in pounds per acre, for designated water and nitrogen levels (Yuma Valley, Arizona, 1970–71)

7.5. Summary of treatment levels and corresponding wheat yields (Yuma Mesa, Arizona, 1971–72)

7.6. Predicted wheat yields, in pounds per acre, for designated water and nitrogen levels (Yuma Mesa, Arizona, 1971–72)

7.7. Summary of treatment levels and corresponding wheat yields (Yuma Mesa, Arizona, 1970–71)

7.8. Predicted wheat yields, in pounds per acre, for designated water and nitrogen levels (Yuma Mesa, Arizona, 1970–71)

7.9. Summary of treatment levels and corresponding wheat yields (Mesa, Arizona, 1971–72)

7.10. Predicted wheat yields, in pounds per acre, for designated water and nitrogen levels (Mesa, Arizona, 1971–72)

7.11. Summary of treatment levels and corresponding wheat yields (Mesa, Arizona, 1970–71)

7.12. Predicted wheat yields, in pounds per acre, for designated water and nitrogen levels (Mesa, Arizona, 1970–71)

7.13. Summary of treatment levels and corresponding wheat yields (Safford, Arizona, 1970–71)

7.14. Predicted wheat yields, in pounds per acre, for designated water and nitrogen levels (Safford, Arizona, 1970–71)

7.15. Summary of treatment levels and corresponding wheat yields (Walsh, Colorado, 1970-71)

7.16. Predicted wheat yields, in pounds per acre, for designated water and nitrogen levels (Walsh, Colorado, 1970-71)

8.1. Summary of treatment levels and corresponding cotton (lint) yields (Shafter, California, 1969)

8.2. Predicted yields of lint cotton, in pounds per acre, for designated water and nitrogen levels (Shafter, California, 1969)

8.3. Summary of treatment levels and corresponding cotton (lint) yields (Shafter, California, 1968)

8.4. Predicted yields of lint cotton, in pounds per acre, for designated water and nitrogen levels (Shafter, California, 1968)

8.5. Summary of treatment levels and corresponding cotton (lint) yields (Shafter, California, 1967)

8.6. Predicted yields of lint cotton, in pounds per acre, for designated water and nitrogen levels (Shafter, California, 1967)

8.7. Predicted yields of lint cotton, in pounds per acre, for designated water and nitrogen levels (Shafter, California, 1967, 1968, and 1969 combined)

8.8. Summary of treatment levels and corresponding cotton (lint) yields (West Side, California, 1969)

8.9. Predicted yields of lint cotton, in pounds per acre, for designated water and nitrogen levels (West Side, California, 1969)

8.10. Summary of treatment levels and corresponding cotton (lint) yields (West Side, California, 1968)

8.11. Predicted yields of lint cotton, in pounds per acre, for designated water and nitrogen levels (West Side, California, 1968)

8.12. Summary of treatment levels and corresponding cotton (lint) yields (West Side, California, 1967)

8.13. Predicted yields of lint cotton, in pounds per acre, for designated water and nitrogen levels (West Side, California, 1967)

8.14. Predicted yields of lint cotton, in pounds per acre, for designated water and nitrogen levels (West Side, California, 1967 and 1969 combined)

8.15. Summary of treatment levels and corresponding cotton (seed) yields (Safford, Arizona, 1971)

8.16. Predicted yields of lint cotton, in pounds per acre, for designated water and nitrogen levels (Safford, Arizona, 1971)

8.17. Summary of treatment levels, corresponding cotton (seed) yields, and percent lint (Tempe, Arizona, 1971)

8.18. Predicted yields of lint cotton, in pounds per acre, for designated water and nitrogen levels (Tempe, Arizona, 1971)

8.19. Summary of treatment levels, corresponding cotton (seed) yields, and percent lint (Yuma Mesa, Arizona, 1971)

8.20. Predicted yields of lint cotton, in pounds per acre, for designated water and nitrogen levels (Yuma Mesa, Arizona, 1971)

8.21. Summary of treatment levels, corresponding cotton (seed) yields, and lint (Yuma Valley, Arizona, 1971)

8.22. Predicted yields of lint cotton, in pounds per acre, for designated water and nitrogen levels (Yuma Valley, Arizona, 1971)

9.1. Summary of treatment levels and corresponding sugar beet root yields, top yields, and sucrose percentages (Mesa, Arizona, 1971-72, May harvest)

9.2. Predicted yields of sugar beet roots at 15 percent sucrose, in tons per acre, for designated water and nitrogen levels (Mesa, Arizona, 1971-72, May harvest)

9.3. Predicted yields of sugar beet tops, in tons per acre, for designated water and nitrogen levels (Mesa, Arizona, 1971-72, May harvest)

9.4. Summary of treatment levels and corresponding sugar beet root yields, top yields, and sucrose percentages (Mesa, Arizona, 1971-72, July harvest)

9.5. Predicted yields of sugar beet roots at 15 percent sucrose, in tons per acre, for designated water and nitrogen levels (Mesa, Arizona, 1971-72, July harvest)

9.6. Predicted yields of sugar beet tops, in tons per acre, for designated water and nitrogen levels (Mesa, Arizona, 1971-72, July harvest)

9.7. Summary of treatment levels and corresponding sugar beet root yields, top yields, and sucrose percentages (Mesa, Arizona, 1970-71, May harvest)

9.8. Predicted yields of sugar beet roots at 15 percent sucrose, in tons per acre, for designated water and nitrogen levels (Mesa, Arizona, 1970-71, May harvest)

9.9. Predicted yields of sugar beet tops, in tons per acre, for designated water and nitrogen levels (Mesa, Arizona, 1970-71, May harvest)

9.10. Summary of treatment levels and corresponding sugar beet root yields, top yields, and sucrose percentages (Mesa, Arizona, 1970-71, July harvest)

9.11. Predicted yields of sugar beet roots adjusted to 15 percent sucrose, in tons per acre, for designated water and nitrogen levels (Mesa, Arizona, 1970-71, July harvest)

9.12. Predicted yields of sugar beet tops, in tons per acre, for designated water and nitrogen levels (Mesa, Arizona, 1970-71, July harvest)

9.13. Summary of treatment levels and corresponding sugar beet root yields, top yields, and sucrose percentages (Mesa, Arizona, 1969-70, May harvest)

9.14. Predicted yields of sugar beet roots adjusted to 15 percent sucrose, in tons per acre, for designated water and nitrogen levels (Mesa, Arizona, 1969-70, May harvest)

9.15. Predicted yields of sugar beet tops, in tons per acre, for designated water and nitrogen levels (Mesa, Arizona, 1969-70, May harvest)

9.16. Summary of treatment levels and corresponding sugar beet root yields, top yields, and sucrose percentages (Mesa, Arizona, 1969-70, July harvest)

9.17. Predicted yields of sugar beet roots adjusted to 15 percent sucrose, in tons per acre, for designated water and nitrogen levels (Mesa, Arizona, 1969-70, July harvest)

9.18. Predicted yields of sugar beet tops, in tons per acre, for designated water and nitrogen levels (Mesa, Arizona, 1969-70, July harvest)

9.19. Predicted yields of sugar beet roots adjusted to 15 percent sucrose, in tons per acre, for designated water and nitrogen levels (Mesa, Arizona, 1970-71 and 1971-72 combined, May harvest)

9.20. Predicted yields of sugar beet tops, in tons per acre, for designated water and nitrogen levels (Mesa, Arizona, 1970-71 and 1971-72 combined, May harvest)

9.21. Predicted yields of sugar beet roots adjusted to 15 percent sucrose, in tons per acre, for designated water and nitrogen levels (Mesa, Arizona, 1970-71 and 1971-72 combined, July harvest)

9.22. Predicted yields of sugar beet tops, in tons per acre, for designated water and nitrogen levels (Mesa, Arizona, 1970-71 and 1971-72 combined, July harvest)

9.23. Summary of treatment levels and corresponding sugar beet root yields, top yields, and sucrose percentages (Yuma Mesa, Arizona, 1970-71)

9.24. Predicted yields of sugar beet roots adjusted to 15 percent sucrose, in tons per acre, for designated water and nitrogen levels (Yuma Mesa, Arizona, 1970-71)

9.25. Predicted yields of sugar beet tops, in tons per acre, for designated water and nitrogen levels (Yuma Mesa, Arizona, 1970-71)

9.26. Summary of treatment levels and corresponding sugar beet root yields, top yields, and sucrose percentages (Yuma Mesa, Arizona, 1969-70)

9.27. Predicted yields of sugar beet roots adjusted to 15 percent sucrose, in tons per acre, for designated water and nitrogen levels (Yuma Mesa, Arizona, 1969-70)

9.28. Predicted yields of sugar beet tops, in tons per acre, for designated water and nitrogen levels (Yuma Mesa, Arizona, 1969-70)

9.29. Summary of treatment levels and corresponding sugar beet root yields, top yields, and sucrose percentages (Yuma Valley, Arizona, 1970-71)

9.30. Predicted yields of sugar beet roots adjusted to 15 percent sucrose, in tons per acre, for designated water and nitrogen levels (Yuma Valley, Arizona, 1970-71)

9.31. Predicted yields of sugar beet tops, in tons per acre, for designated water and nitrogen levels (Yuma Valley, Arizona, 1970-71)

9.32. Summary of treatment levels and corresponding sugar beet root yields, top yields, and sucrose percentages (Yuma Valley, Arizona, 1969-70)

9.33. Predicted yields of sugar beet roots adjusted to 15 percent sucrose, in tons per acre, for designated water and nitrogen levels (Yuma Valley, Arizona, 1969-70)

9.34. Predicted yields of sugar beet tops, in tons per acre, for designated water and nitrogen levels (Yuma Valley, Arizona, 1969-70)

9.35. Summary of treatment levels and corresponding sugar beet root yields and sucrose percentages (Safford, Arizona, 1970)

9.36. Predicted yields of sugar beet roots, in tons per acre, for designated water and nitrogen levels (Safford, Arizona, 1970)

9.37. Summary of treatment levels and corresponding sugar beet root yields and sucrose percentages (Fort Collins, Colorado, 1969)

9.38. Predicted yields of sugar beet roots, in tons per acre, for designated water and nitrogen levels (Fort Collins, Colorado, 1969)

9.39. Summary of treatment levels and corresponding sugar beet root yields and sucrose percentages (Walsh, Colorado, 1970)

9.40. Predicted yields of sugar beet roots, in tons per acre, for designated water and nitrogen levels (Walsh, Colorado, 1970)

9.41. Irrigation dates and quantities of water applied in acre-inches for five irrigations in production of sugar beet roots (Plainview, Texas, 1971)

9.42. Summary of treatment levels and corresponding sugar beet root yields and sucrose percentages (Plainview, Texas, 1971)

9.43. Predicted yields of sugar beet roots, in tons per acre, for designated water and nitrogen levels (Plainview, Texas, 1971)

9.44. Sensitivity analysis of canonical correlations for different sizes of elasticities

10.1. Distribution of percent sand throughout specified soil profile, estimated weights of relative importance of soil depth on plant growth and yield, and weighted average of percent sand for 1971 wheat and 1971-72 sugar beet experiments at Mesa, Arizona

10.2. Soil-engineering classes for four soil characteristics as specified by the Soil Conservation Service and coded values assigned to each class

10.3. Predicted yields and 95% confidence limits for predicted yields derived from generalized production function for corn grain compared with observed yields for specified treatment combinations and experimental sites, in pounds per acre

10.4. Predicted yields and 95% confidence limits for predicted yields derived from generalized production function for corn silage compared with observed yields for specified treatment combinations and experimental sites, in pounds per acre

10.5. Predicted wheat yields, 95% confidence limits for predicted yields, and observed yields for specified treatment combinations and experimental sites, in pounds per acre

10.6. Predicted cotton (lint) yields, 95% confidence limits on predicted yields, and observed yields for specified treatment combinations and experimental sites, in pounds per acre

10.7. Predicted root yields (15% sucrose), 95% confidence limits on predicted yields, and observed yields for specified treatment combinations and experimental sites, in tons per acre

11.1. Quantities of water demanded for producing corn when corn price is $0.04 per pound and three levels of water price and nitrogen are assumed (Colby, Kansas, 1971)

11.2. Quantities of water demanded for producing wheat when wheat price is $0.06 per pound and three levels of water price and nitrogen are assumed (Yuma Valley, Arizona, 1971-72)

11.3. Quantities of water demanded for producing cotton when lint price is $0.30 per pound and three levels of water price and nitrogen are assumed (West Side, California, 1969)

11.4. Quantities of water demanded for producing sugar beets when beet price is $18 per ton and three levels of water price and nitrogen are assumed (Mesa, Arizona, 1971–72)

12.1. Quantity of corn (grain) produced, in pounds per acre, when P_w is $1 per acre-foot and P_y and N are at the specified levels (Colby, Kansas, 1971)

12.2. Quantity of wheat produced, in pounds per acre, when P_w is $1 per acre-foot and P_y and N are at the specified levels (Yuma Valley, Arizona, 1971–72)

12.3. Quantity of lint cotton produced, in pounds per acre, when P_w is $1 per acre-foot and P_y and N are at the specified levels (West Side, California, 1969)

12.4. Quantity of sugar beets produced, in tons per acre, when P_w is $1 per acre-foot and P_y and N are at the specified levels (Mesa, Arizona, 1971–72, July harvest)

13.1. Average monthly flows during 1920–59 in Huntington and Cottonwood watersheds and derived monthly water requirements, in 1000 acre-feet

13.2. Optimal production patterns for medium-sized beef (BF) farm in the Cottonwood watershed with high and low prices for irrigation water

13.3. Water shortages and corresponding marginal value products (MVP) simulated for "preproject" and "project" conditions (1920–59, Huntington and Cottonwood areas)

Appendix Table 1. Estimated relative importance of indicated soil layer on crop yield for specified crops and locations

Appendix Table 2. Summary of soil characteristics weighted according to weights in Appendix Table 1 for specified crops and experimental sites

Appendix Table 3. Planting and harvesting dates and estimates of "critical" crop growth periods for specified crops and locations

WATER PRODUCTION FUNCTIONS FOR IRRIGATED AGRICULTURE

ROLE OF RESPONSE FUNCTIONS IN MANAGEMENT
AND DECISIONS ON WATER USE

Knowledge of water response functions is an important set of the information needed in either private or public decision on optimal water programs. Unfortunately, however, yield response functions for water have seldom been known before large irrigation projects have been initiated from surface waters or private farmers have developed supplies of groundwater. The statement applies in the United States and all other countries of the world.

A considerable number of water response functions have been estimated in this study for a range of locations, soil, and environmental conditions. Knowledge is broadened accordingly and generalized functions are derived with some success. The task is, however, far from complete. The project upon which the estimates are forthcoming extended for only three years under a limited budget provided by the Bureau of Reclamation. For full generalization of water response or production functions, so that they can be used to predict yields under a range of soil and climatic conditions, a large-scale project is needed, which extends over great space, many years, and a wide range of measured environmental variables. And whereas the current project had a modest budget of thousands of dollars, the budget needed for comprehensive accomplishments must stretch far into the millions of dollars. Projects of this financial magnitude are now being financed by the Agency for International Development through Colorado State University, Utah State University, and other institutions. Experiments are being conducted under a range of soil and climatic conditions in several countries of South America and Southeast Asia.

While the project under discussion was modest in time and available resources, it did meet with some initial successes that provide a foundation of optimism in projecting hopes of accomplishments for research of the future less restrained by resources, manpower, and time. Its success was due not alone to the administrators and professionals of the Bureau of Reclamation and Iowa State University but also to the agronomists, engineers, and agricultural economists in several western states who cooperated vigorously in implementation of the project. The fact that the experiments had to be conducted with facilities and at stations in locations already developed and available to the cooperating scientists limited the "sample in space" relative to the soil and environmental conditions or variables that could be incorporated into the model. For future experiments having more funds and time available and lacking these rigidities, it should be possible to establish water response experiments so that they cover a more complete range and magnitude of variables that adequately represent soil and climatic variables.

THE ROLE OF PRODUCTION FUNCTIONS IN
WATER RESOURCE PLANNING

In most irrigation projects, the available supply of water is limited and can be used on a number of soil types or land groups within transportation distance from the supply. The productivity of water will differ among the different soils and it will differ when it is applied at different levels. Finally, yield from water will interact with fertilizer and other management inputs— as well as the soil and climatic variables. Often the supply of water will be stochastic or vary from year to year. The stochastic characteristics may arise because of variations in snow pack or rainfall that serve as the origin of the water supply. Decision rules for optimal water use thus will depend on (a) the knowledge of the water production function relative to various soils, environmental variables, and management variables with which it can be used and (b) the stochastic nature of the water supply. Generally in this study the stochastic characteristics of water supplies are not considered, although one phase of the project did include development of a decision model for varying water supplies.[1] Also, estimation of water production functions is the major goal. Some illustrative applications of the predicted functions are made but these major uses are left for other studies. For example, in another phase of the project, the experimental data were used in linear programming models to estimate optimal water use on individual farms and to derive water demand for irrigation projects.[2]

SIMPLE PROJECT ILLUSTRATION

While the estimated water response functions generally have not been available, many and large irrigation projects have been built over the world that specify water distribution patterns among soils groups and farmers. Sometimes these distributions are based on equity considerations since the added water resources increase the value of the farmer's other resources and increase his income. Hence, methods are sometimes devised, apart from the marginal productivity of water, to transfer equitable shares of water to individual farmers, villages, or communities. Water rights that become historic resource properties sometimes arise from these procedures. In other cases, legal water rights are only historic in nature and relate but little to water productivity.

Leaving aside both equity and historic rights considerations, we can illustrate simply the importance of water response knowledge in the development of water projects. Suppose, for example, that the total supply of water available is W_T and that it can be used on four soil groups or land classes. Each of these land classes has a water response or production function as in (1.1) through (1.4) where Y_i is yield on the ith land class and W_i is the amount of water used on the ith land class.

$$Y_1 = a_1 W_1 - b_1 W_1^2 \tag{1.1}$$

$$Y_2 = a_2 W_2 - b_2 W_2^2 \tag{1.2}$$

$$Y_3 = a_3 W_3 - b_3 W_3^2 \qquad (1.3)$$

$$Y_4 = a_4 W_4 - b_4 W_4^2 \qquad (1.4)$$

The W_T water supply available will maximize the forthcoming product of a given crop if it is allocated to each land class so that the marginal productivity of water (the last increment used on each) is equal for the four land areas. Hence, we must equate the marginal productivity of water on each land class but at an unknown level. Thus the task is to simultaneously solve for the amount of water and the marginal productivity of water on the four land areas. Assigning the unknown marginal productivity of water the value m, we can set the marginal productivity of water on each land area, the derivative of yield with respect to water supplied, equal to m as in the equation set (1.5) where

$$\frac{dY_1}{dW_1} = a_1 - 2b_1 W_1 = m$$

$$\frac{dY_2}{dW_2} = a_2 - 2b_2 W_2 = m$$

$$\frac{dY_3}{dW_3} = a_3 - 2b_3 W_3 = m \qquad (1.5)$$

$$\frac{dY_4}{dW_4} = a_4 - 2b_4 W_4 = m$$

$$W_1 + W_2 + W_3 + W_4 = W_T$$

We also have added the restraint or requirement that the total amount of water used on the four land areas is equal to the total water supply available. We now have five equations in five unknowns (W_1, W_2, W_3, W_4, and m). Rearranging terms with the known parameters on the right-hand side and the unknown variables and their coefficients on the left-hand side, we have the matrix relationship in (1.6).

$$\begin{bmatrix} 2b_1 & 0 & 0 & 0 & 1 \\ 0 & 2b_2 & 0 & 0 & 1 \\ 0 & 0 & 2b_3 & 0 & 1 \\ 0 & 0 & 0 & 2b_4 & 1 \\ 1 & 1 & 1 & 1 & 0 \end{bmatrix} \begin{bmatrix} W_1 \\ W_2 \\ W_3 \\ W_4 \\ m \end{bmatrix} = \begin{bmatrix} a_1 \\ a_2 \\ a_3 \\ a_4 \\ W_T \end{bmatrix} \qquad (1.6)$$

Solving for the values of W_i and m through the inversion of the square matrix on the left ($w = C^{-1}a$ where w is the left-hand vector of water and m variables,

C is the 5×5 matrix at the left, and a is the vector or right-hand side of known parameters), we can determine the amount of water optimally to be allocated to each of the four land areas if production from the given water supply, W_T, is to be at a maximum. Of course, with a maximum output the marginal productivity of water is the same on all four land areas, even though they may differ in productivity, and their water response functions illustrated in Equations (1.1) through (1.4) are extremely simple, and certainly more detailed and sophisticated models of water allocation can be specified. However, to our knowledge, even such simple response functions have not been available in the development of any major water project over the world, nor have even simple allocative models such as the above been used in water distribution schemes.

SIMPLE RESPONSE FUNCTIONS

While such simple response functions have not generally been available, this study presents the analysis of water production functions estimated in several western states of the nation. The functions estimated are of rather simple nature, although they do extend knowledge of water response far beyond that of previous information. Much additional research must be completed before the great paucity of water response information is overcome, however. We do extend estimation beyond the simple response equations estimated above. As is clearly known both for developed and developing countries, water interacts strongly with other management inputs such as fertilizer in increasing yields. Hence, production functions estimated in this study incorporate different levels of fertilizer as an experimental variable. Finally, since response to both irrigation water and fertilizer is expected to vary with rainfall and its seasonal distribution, daylight hours, temperature, available soil moisture, soil characteristics, and other environmental conditions of particular locations or sites, variables representing certain of these conditions also are incorporated into the response functions. While a number of soil and environmental variables were specified for measurement when the various experiments were initiated, only a limited number proved statistically significant. However, these efforts were of sufficient success to encourage further efforts in estimating generalized production functions. Larger projects with more funds, time, and experimental locations should result in improved generalized functions that incorporate more of the relevant soil and environmental variables. These extensions also may need to consider plant root distribution and moisture extraction as affected by genetic plant characteristics as well as soil and climatic variables.

The goal in incorporating soil and environmental variables or characteristics is to allow prediction of yield response to irrigation level over a wider range of soil, climate, and locational conditions. Attainment of this goal would eliminate the necessity of conducting an experiment on each particular soil type or association in order that productivity of water in interaction with other variables might be known both for farm planning and project evaluation.

DYNAMICS OF WATER RESPONSE

The water production functions estimated in this study also are simple in the sense that they do not consider or vary time of water application. Effectively, the different levels of water are used as treatments of the same time periods. However, a considerable difference in response from an acre-foot of water might be expected depending on whether it is all applied at planting time or midway in the plant growth cycle. Also, different responses would be expected if the given amount of water was applied in four or eight equal amounts at different times in seed bed preparation or plant growth. Other variations in time and method of application also will alter the expected yield response. These aspects of the dynamics of water use and response await larger and (or) more complex experiments than were allowed in this study.

Time of irrigation and plant growth interact with other moisture and soil characteristics to affect plant stress, wilting, and yield. Beringer indicates that, even within one layer of soil, moisture tension will be an increasing function of time and, in any one period of time, moisture tension within the soil profile, between layers, will be a decreasing function of soil depth.[3] Wadleigh[4] and Taylor[5] suggested methods for combining these distributions into a single number by means of an integrated moisture stress concept. By Wadleigh's method, total soil moisture stress at any point in time is a linear function of moisture tension and the osmotic pressure of salts in the soil. Soil moisture tension is a decreasing function of depth, and osmotic pressure is an increasing function of depth due to a greater accumulation of salts in lower soil strata. To the extent that these two effects cancel each other, moisture stress as it affects plant growth then is no longer a function of depth but is a function of time. By Taylor's method, variation in the integrated moisture stress in the root zone that results from more frequent irrigation will be partly a function of labor and capital inputs as well as water quantity up to some specified level.

If the soil was at field capacity at the time plant growth began and was kept at capacity, the maximum rate of growth presumably would occur. However, if a given amount of water is applied at this time, with decision to apply water again at a later time, the question is then twofold: How much more water should be applied and when should it be applied? The answer to both questions will depend on the amount of water applied (or the soil moisture available) at the outset in relation to the moisture-holding capacity of the soil. Given the initial moisture or irrigation regime, crop response to water will depend on when water is again applied, how much is applied, and how much time elapses in the growing season until a third application is made. Thus the complex of responses to water application might appear as in Figure 1.1.

Initial water application at different levels at time t_0 alone might result in the response curve a. If a second set of applications could be made at t_1, an added yield response of c may result; although if they were made at a later time, t_2, only a slightly greater response, b, may be realized. Similarly, if a

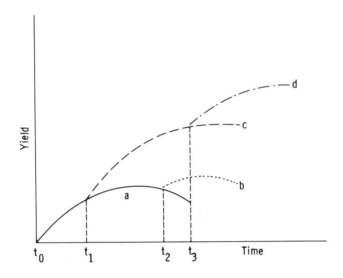

FIG. 1.1. Dynamics of crop yield response to time and amount of irrigation.

third set of irrigations is applied at t_3, the further response d may be realized. The times indicated refer only to the time of each irrigation. Yield response relates to the amount of water that could be added at each point in time (but which are not represented on the graph). The response functions refer to irrigation levels at a given point in time (and not to different expanses of time).

Obviously, then, water response functions are complex functions (a) in their dynamic nature, (b) in their interactions with other biological inputs such as fertilizer, plant variety, pesticides, etc., and (c) in conformance with their surrounding soil and climatic environment. This monograph deals with the second and third considerations above, but does not treat the dynamics of water response. However, two studies serving as part of the overall project upon which the following estimates are made do treat time, stochastic aspects of water supply, and decision procedures as they relate to the dynamics and time aspects of water supplies and response.[6]

PRODUCTION FUNCTIONS AND ECONOMIC APPLICATIONS

Production functions are used in many economic analyses. Any discussion of commodity supply, demand for inputs, and income distribution is predicated on some knowledge of input-output relationships. This knowledge, however, tends to be incomplete, particularly for commodities involving several inputs. Several input-output relationships are unknown. Others are currently unmeasurable or have not been measured. Consequently, the extent to which production functions can be estimated to approximate actual input-output relationships depends on knowledge of these relationships and data availability.

Analyses of production functions exist on other levels. Conceptual and theoretical formulations can provide hypotheses to be tested and can be used to identify gaps in data availability. These latter considerations are important in devising appropriate survey and experimental designs for generating necessary data.

In a more formal vein, a production function represents a schedule or mathematical formulation expressing the relationships between inputs and outputs. It also indicates the maximum amount of product obtainable from a specified quantity of inputs given the existing technology governing the input-output relationships. By definition, a production function embodies technical efficiency. This requires that a specified set of inputs cannot be recombined to produce a larger output or a specific level of output cannot be produced with fewer inputs. Consequently, production should occur on the "frontier" of a production possibility surface, suggested through curve OA in Figure 2.1. In this simplified example, assume the input-output relationship is known with certainty so that OX_1 of water applied to an unspecified number of other inputs having fixed use-levels is expected to produce OY_2 of output rather than, for example, OY_1. A producer planning to operate along OA in Figure 2.1 must, however, not only be aware of this technology but have access to it and be able to successfully adopt it. Improvements on existing technology may permit him to move to OB where each input use-level results in output levels above those realizable with OA. Alternatively, OY_2, for example, can be generated with OX_1', thereby freeing $X_1'X_1$ units of water for use elsewhere. As with OA, the input-output relationships implicit in OB are assumed to be known with certainty. Since these relationships are neither fully known nor controllable, a distribution of yields would be associated with each input use-level. This range of expected yields depends on the estimated variability of the predicted yield corresponding to the specified input use-level. This concept is discussed in Chapter 4. Finally, inputs included in a production function are assumed to be homogenous. The chemical compo-

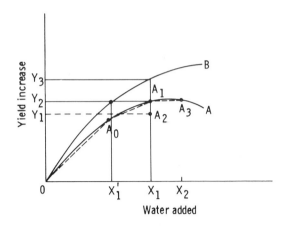

FIG. 2.1. Hypothetical production functions (response surfaces) for yield-water relationships.

sition of OX_1 units of water in Figure 2.1 is assumed to be the same as that for the $X_1 X_2$ units.

PLAN FOR CHAPTER 2

Following a discussion of the nature of a single variable input production function, multivariate and multiple functions are introduced. Because of the computational ease, several of the concepts developed are based on Cobb-Douglas production functions. The properties of these functions are discussed in Chapter 3. Several of the concepts presented in this chapter are quantified in subsequent chapters.

POINT ESTIMATES AS COMPARED TO CONTINUOUS PRODUCTION FUNCTIONS

Points such as A_1 and A_0 in Figure 2.1 represent point estimates of yield-water relationships. If sufficient points are available, including A_1, a continuous production function, OA, can be derived. Sufficient data are often not available for estimating continuous functions. Point estimates are sometimes obtained in field surveys where respondents indicate, for example, the quantity of water applied and the corresponding per acre yield. If several point estimates are available, a segmented linear approximation of the continuous relationship may be possible. The segmented curve $OA_0 A_1 A_3$ in Figure 2.1 could be used as an approximation of OA. If points A_0, A_1, and A_3 are based on cross-section data, that is, data from three different production units, the segmented input-output relationship could result in a confounding of additional factors such as differences in soil, production techniques, and managerial efficiency. In this situation, certain production inputs are no longer homogenous.

The limitations associated with single-point estimates are fairly straight-forward. As price relationships for inputs and outputs vary, producers attempting to maximize profits must usually change the mix of inputs. In Figure 2.1, the producer should move from, for example, A_1 to another point on OA. With knowledge of the response surface being limited to A_1, the position and slope of points to the right and left of A_1 are not known. Consequently, the appropriate use-level for the input cannot be determined. Point estimates, however, are better than no estimates.

Field experiments are often used to generate sufficient data for estimating continuous production functions. When two or more inputs such as water and fertilizer are incorporated into the production function, marginal rates of substitution between inputs for producing a specified level of output can be estimated. Continuous production functions also permit estimation of other concepts such as isoquants, isoclines, production elasticities, and product supply functions. These concepts discussed later in this chapter are not derivable from single-point estimates.

PRODUCTION FUNCTIONS WITH ONE VARIABLE INPUT

A single variable production function, as specified in (2.1), is of little practical significance. Let Y denote output, X_1 the variable input, and f^* the input-output relationship. Few, if any, actual production relationships involve a single input. When (2.1) is rewritten as (2.2), a more meaningful relationship is expressed.

$$Y = f^*(X_1) \tag{2.1}$$

$$Y = f(X_1 | X_2, X_3, X_4) \tag{2.2}$$

Let Y represent per acre crop yield, X_1 equal acre-inches of water applied, X_2 equal pounds per acre of fertilizer applied, X_3 equal plant population, and X_4 equal other inputs available in a fixed quantity. The functional relationship between Y and the X's is represented by f as distinguished from f^* in (2.1). The vertical bar in (2.2) denotes X_1 as the only variable input applied to specified, fixed levels of X_2, X_3, and X_4. Assume X_2 is 98 pounds of fertilizer per acre, X_3 is 20,000 plants per acre, and X_4 represents 10 hours of labor. Let k represent the combined effects of X_2, X_3, and X_4 and assume f represents a Cobb-Douglas function, as in (2.3), with the corresponding predicted yield equation.

$$Y = kX_1^{b_1} \tag{2.3}$$

$$\hat{Y} = 25 X_1^{.3} \tag{2.4}$$

These hypothetical relationships are depicted in Figure 2.2 where k_1 and k_2 represent the effect of 98 and 250 pounds of fertilizer per acre, respectively,

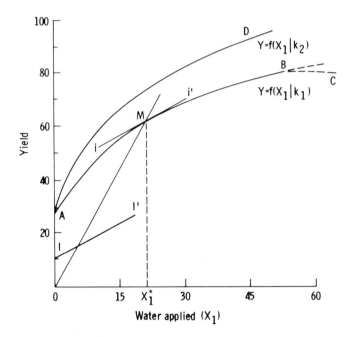

FIG. 2.2. Yield response to water applied at two levels
of fertilization.

with X_3 and X_4 invariant. Output OA is generated by available soil moisture
and precipitation. Assume the producer is operating along response curve AB
in Figure 2.2.   Several rather basic physical and economic concepts are
derivable from this relationship. These concepts are important components
of the economic analyses of water use.

Average and marginal products
      The average product (AP) of an input, for example, X_1 in Figure 2.2., is
defined as the total product (yield) divided by the amount of input required
to produce that output, as in (2.5a).

$$AP = Y/X_1 = f(X_1 | X_2, X_3, X_4)/X_1 \qquad (2.5a)$$

      Using (2.4) to derive AP, we have (2.5b) where the average product

$$AP = 25X_1^{.3}/X_1 = 25X_1^{-.7} = 25/X_1^{.7} \qquad (2.5b)$$

is declining with greater values of input X_1.   The AP declines for total response or production curves AD and AB in Figure 2.2. When the level of X_1
used is X_1^* in Figure 2.2, the AP is the equivalent of the slope of the line OM
that intersects the origin and the point of the total product curve that con-
forms with X_1^* amount of input applied.

The marginal physical product, *MP*, of an input is the addition to total product resulting from use of one more unit of the input, other resources held constant in level. More exactly, the marginal product is the first derivative of output with respect to the particular input or

$$MP = \frac{dY}{dX_1} \tag{2.6a}$$

For the general production function in (2.2) the marginal physical product for X_1 is

$$MP_1 = \frac{dY}{dX_1} = df(X_1 \mid X_2, X_3, X_4)/dX_1 \tag{2.6b}$$

The marginal product corresponding to the form of the production function in (2.4) is

$$MP_1 = 7.5 X_1^{-.7} = \frac{dY}{dX_1} \tag{2.6c}$$

Graphically, the marginal physical product is the slope of the total product or yield function. Thus in Figure 2.2. the marginal product of input X_1 used at X_1^* level is the slope of line ii' tangent to AB at point M. The MP of input X_1 is positive but steadily declines from point A along AB. For a yield response represented by AMC, the MP of X_1 is negative beyond point B (that is, the first derivative of output with respect to X_1 is negative) and output declines as more of the input is used. To use X_1 in excess of 50 units (conforming to point B on the response curve) would be uneconomic.

Economically optimum input use-level

Referring to the simplified example in Figure 2.2, assume the producer's objective is to maximize net returns over variable costs. In this situation, additional units of X_1 can be profitably used as long as the addition to total revenue exceeds the addition to variable costs. This objective is represented by the profit function in (2.7) where π represents profit and P_y and P_1 are the per unit revenue (price) and cost of the output and input, respectively.

$$\pi = P_y Y - P_1 X_1 = P_y f(X_1 \mid X_2, X_3, X_4) - P_1 X_1 \tag{2.7}$$

Profits to fixed resources are maximized when the derivative of profit, π, with respect to the input, X_1, is set to zero. Hence, for this condition relative to the general profit function in (2.7), we have (2.8).

$$\frac{d\pi}{dX_1} = P_y \frac{dY}{dX_1} - P_1 = 0 \tag{2.8a}$$

By solving (2.8a) for the X_1, we obtain the amount of input that will maximize profits. For a numerical example, we use the production or response function in (2.4). Multiplying it by the price of Y (since the function is the value of X) and subtracting the product of the input multiplied by the price, we have

$$\pi = P_y 25 X_1^3 - P_1 X_1 \qquad (2.8b)$$

where the marginal profit, or the profit-maximizing condition, is

$$\frac{d\pi}{dX_1} = 7.5 P_y X_1^{-.7} - P_1 = 0 \qquad (2.8c)$$

Solving for X_1, $X_1 = (7.5 P_y P_1^{-1})^{1.4286}$, we have the amount of input or water that will maximize profit per acre (if fixed resources refer to an acre of land). If $P_y = 1.50$ and $P_1 = 1.35$, the value of X_1 that will maximize profit per acre is 20.7 acre-inches and yield or Y is approximately 62.

Returning to the condition for profit maximization in (2.8a), we can simplify to

$$MP_1 = \frac{dY}{dX_1} = P_1/P_y \qquad (2.9)$$

which indicates that for profit maximization the marginal physical product of the input, MP, must be equal to the price ratio represented by input price divided by output price. This condition is equivalent to superimposing the relevant price ratio on curve AB in Figure 2.2, thereby generating a point of tangency at point M. Since the price ratio line, ii', is tangent to AB, the slopes of AB and ii' are equal at M. In other words, the price ratio is equal to the marginal product of X_1 when it is used at the level X_1^*.

The profit-maximizing criterion in (2.9) is based on several important assumptions. First, X_1 is the only variable input being analyzed in the production of a single output and the production function is concave to the X_1 axis. Second, the nature of the input-output relationship is known so that MP_1 can be estimated. P_y and P_1 are also known and invariant. Sufficient X_1 is available to realize the equality in (2.9); that is, there are no constraints on resource availability. Finally, the producer's objective is profit maximization and his capital is unlimited. If these assumptions and conditions cannot be satisfied, (2.9) must be modified accordingly. The maximizing principle implicit in (2.9) is generalizable to include several inputs and outputs and objectives other than profit maximization.

One additional comment on Figure 2.2 is of interest. By rewriting (2.7) and substituting the prices previously assumed, an iso-profit line can be derived as in (2.10).

$$Y = \pi/P_y + (P_1/P_y)X_1$$
$$= \pi/1.50 + .9 X_1 \qquad (2.10a)$$

If π is arbitrarily set at \$15 per acre, then

$$Y = 10 + .9 X_1 \qquad (2.10b)$$

if $X_1 = 0$, $Y = 10$; if $X_1 = 10$, $Y = 19$. Using these relationships, the iso-profit function in (2.10) is plotted as line II' in Figure 2.2. By assuming different values of π but keeping P_y and P_1 constant, a family of parallel iso-profit lines is derived. The producer planning to maximize profits attempts to reach the highest attainable iso-profit line. For the producer operating along AB in Figure 2.2, the highest iso-profit line is the one tangent to curve AB, namely ii'

Static demand for a single input

Knowledge of the production function also allows derivation of the static demand function for an input such as water. This demand function is represented as the marginal value product of the resource input. The marginal value product, MVP, is computed by multiplying the marginal physical product, MP, by the price of the product. Hence $MVP = P_y \cdot MP$ or $MVP = P_y dY/dX_1$. Profit maximization also can be defined relative to the marginal value productivity of an input or resource. Profit will be maximized when the marginal value product is equal to the price of the input, $MVP_1 = P_1$. This is the same condition stated in (2.8a) since the term $P_y \cdot dY/dX_1$ is the marginal value product. If the price of the input is transposed to the right-hand side of the equal sign, the marginal value productivity of X_1 is equated to the price of the input, X_1. If P_1 above, the price of X_1 or water, increases while all other variables remain fixed, the $P_y MP_1$, the marginal value product of water, becomes less than the price of water. With P_y invariant, equality of the marginal value product with the price of water is restored by using less of X_1 so that its marginal physical product, MP_1, is increased. Thus, as P_1 is varied while P_y is constant, corresponding values of X_1 are defined which maximize profits. These alternative price-quantity combinations (P_1 and X_1) represent points on a static demand curve for the input. As derived in a later chapter, this schedule is static since demand relates to a point in time. Production also is assumed to occur in the area of the production surface where both the marginal product and the average product are positive but diminishing.[1] Using the profit-maximizing condition in (2.8a) applied to the production function in (2.3), the demand function for X_1 is derived as

$$X_1 = [b_1 k P_y P_1^{-1}]^{1/(1-b_1)} \tag{2.11}$$

$$X_1 = [7.5 \, P_y P_1^{-1}]^{1.429} \tag{2.12}$$

The quantity of X_1 demanded varies directly with b_1 and k, the parameters defining the technical input-output relationships in (2.3). If technological innovations raise the values of the technical coefficients of the inputs b_1 and (or) k, if P_y increases, or if the use-level of other inputs increases, greater amounts of X_1 can be used profitably. The quantitative impact of changes in b_1, for example, is represented by the first derivative of (2.11) with respect to b_1. The level of X_1 varies inversely with P_1. Using the numerical quantities

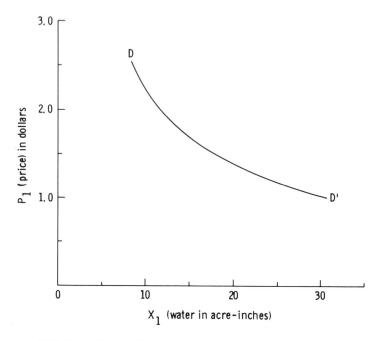

FIG. 2.3. Demand curve for water when X_2 (fertilizer) is
98 pounds per acre.

of production function (2.4), (2.11) is transformed into (2.12). With P_y and
P_1 at 1.50 and 1.35, respectively, X_1 is derived, as shown previously, as about
20.7 acre-inches. This combination of $P_1 = 1.35$ and $X_1 = 20.7$ represents
one point on the derived demand schedule for X_1 in Figure 2.3. By assuming
different values for P_1 in (2.12), the corresponding X_1 in Figure 2.3. is derived. Also if P_y, k, or b_1 increase, DD' shifts to the right. DD' can be veri-
fied as an MVP curve for X_1. Consequently, using the relationship in (2.8),
the intersection of P_1 with DD' would denote the profit-maximizing use-level
for X_1. The assumptions and limitations specified for deriving the economically optimum use-level for an input also apply here.

Price elasticity of demand

In continuing the discussion of the relationship between X_1 and P_1, the
price elasticity of demand for X_1 is a coefficient of the responsiveness in
quantity of X_1 demanded as P_1 varies. This coefficient is defined as the percent change in quantity demanded divided by the associated percent change
in price. In other words, the price elasticity coefficient is the derivative of X_1
with respect to P_1 multiplied by the ratio P_1/X_1 for the price of the input.
Relative to the price of the output, P_y, the demand elasticity is the derivative
of X_1 with respect to P_y multiplied by the ratio P_y/X_1. These elasticities

derived from (2.11) are expressed as

$$\epsilon_{X_1 P_1} = (dX_1/dP_1)(P_1/X_1) = -1(1 - b_1) \quad (2.13)$$

$$\epsilon_{X_1 P_y} = (dX_1/dP_y)(P_y/X_1) = 1/(1 - b_1) \quad (2.14)$$

The negative sign in (2.13) denotes that if b_1 is less than 1.0, prices and quantities demanded of X_1 should move in opposite directions. With b_1 less than 1.0, the price elasticity of demand with respect to P_y is positive, and increases in P_y should increase the use of X_1.

To this point, only one output and a single variable input have been considered in a static context. This is a grossly simplified situation. Technical efficiency has been assumed as well as the availability of sufficient capital to acquire X_1 and to operate at any point on the response curve being considered. Prices and technical coefficients are assumed to be known with certainty, and the producer's assumed goal is profit maximization. The use of this simplified framework does, however, permit the isolation and examination of some of the economic constructs that individuals implicitly or explicitly use in more highly complicated decision-making situations. As will be apparent subsequently, decision-making criteria become increasingly complex as additional inputs and outputs are included in the analyses.

Production possibilities curve

Most inputs can be used in producing alternative products. All available inputs can be committed to a single product, or they can be allocated to producing two or more products. For convenience, assume a single variable input is used to produce either of two output products. A production possibilities curve represents alternative combinations of two outputs that can be produced with a fixed total quantity of that input.

Assume two production functions similar to (2.3) where Y and Z correspond, for example, to corn and soybeans produced with variable input X_1.

$$Y = f(X_{1y}) \quad (2.15a)$$

$$Z = g(X_{1z}) \quad (2.15b)$$

Further assume that a fixed quantity of X_1, denoted by X_1^0, is available.

$$X_1^0 \geqq X_{1y} + X_{1z} \quad (2.16)$$

According to (2.16), X_{1y} and X_{1z} are homogenous inputs whose sum cannot exceed X_1^0. Equations (2.15) and (2.16) can be rewritten so that X_1 is a function of Y and Z.

$$X_{1y} = f^*(Y) \quad (2.17a)$$

$$X_{1z} = g^*(Z) \quad (2.17b)$$

In (2.17), for example, various levels of Y can be considered and the corresponding input requirements, X_{1y}, can be derived as shown in (2.17a). The same relationship for Z is given in (2.17b). Note that the functional form has changed from f to $f*$. With X_{1y} and X_{1z} so expressed, we can designate X_1^0 as

$$X_1^0 \gtreqqless f*(Y) + g*(Z) \tag{2.18}$$

Taking the total differential of (2.18), the alternative combinations of Y and Z that can be produced with X_1^0 are defined as

$$dX_1^0 = (\partial f*/\partial Y)dY + (\partial g*/\partial Z)dZ = 0 \tag{2.19}$$

Since the availability of X_1 is fixed at X_1^0, $dX_1^0 = 0$. By transposing either the Y or Z terms to the right, increases in the production of Y are associated with decreases in the output levels of Z and vice versa. This trade-off is depicted in Figure 2.4. At point A, for example, X_1^0 can be used to produce 50 units of Y and 39 units of Z.

When a portion of X_1^0 is shifted from production of Y to Z or from Z to Y, the trade-off is termed the rate of product transformation, RPT (that is, the marginal rate of substitution of one product for the other). The expression for RPT in (2.20) is derived using the implicit function rule for differentiating equations.

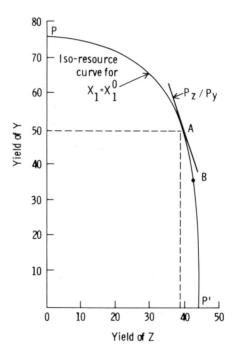

FIG. 2.4. Production possibilities curve for X_1^0 in production of Y and Z.

$$RPT = MP_{1y}/MP_{1z} = -(dY/dZ) \tag{2.20}$$

According to (2.20), the RPT is also equal to the ratio of the marginal physical products of X_1 used in the production of Y and Z, respectively. Assume $Y = 25\,X_1^3$, $Z = 10\,X_1^4$, and $X_1^0 = 40$ with (2.18) rewritten as

$$40 \geqq (.04\,Y)^{3.33} + (.1Z)^{2.5} \tag{2.21}$$

and

$$RPT = 7.5\,(10.1)^{-.7}/4\,(29.9)^{-.6} = -2.9 \tag{2.22}$$

If Y is set at 50, 10.1 units of X_1^0 are required, leaving 29.9 units to produce about 39 units of Z. These values for Y and Z correspond to point A in Figure 2.4 where $dY/dZ = -2.9$. That is, if one additional unit of Z is produced, 2.9 units of Y must be foregone. If a price ratio, P_zP_y, is superimposed and is tangent at point A, $Y = 50$ and $Z = 39$ represent the output levels that maximize total revenue. At point A, $P_z/P_y = -dY/dZ$, or the price ratio is equal to the marginal rate of substitution of product Z for product Y. If P_z increases relative to P_y so that the price ratio is tangent at, for example, point B, a different optimal output configuration results. Compared to point A, a portion of X_1 should be shifted from Y to production of Z.

This simplified example is generalizable to several inputs producing several outputs. Since graphic analysis is best suited for situations requiring two- or three-dimensional space, multivariable, multiple output frameworks are best represented by a system of equations.

MULTIVARIABLE PRODUCTION FUNCTION

If function (2.2) is modified to incorporate a second variable input, as in (2.23), additional physical and economic relationships affecting producers' decision-making processes or procedures for optimum resource use can be derived.

$$Y = h\,(X_1, X_2 | X_3, X_4) \tag{2.23}$$

Recall that X_1 and X_2 represent acre-inches of water and pounds of fertilizer applied per acre, respectively. Note that the unspecified input-output relationship has been changed from f in the one variable input function to h. If (2.23) is quantified as in (2.24), a three-dimensional response surface such as Figure 2.5 can be generated. When $X_2 = 100$, curve AB in Figure 2.5 delineates a water response function corresponding to AB in Figure 2.2. When $X_1 = 0$, for example, OC represents a fertilizer response curve. Other curves originating along the X_1 or X_2 axes are similarly interpreted. The corresponding marginal and average products for X_1 and X_2 are derived according to Equations (2.5) and (2.6).

Product isoquants

In addition to reproducing input-output relationships, product isoquants can be superimposed on Figure 2.5. A product isoquant shows all possible

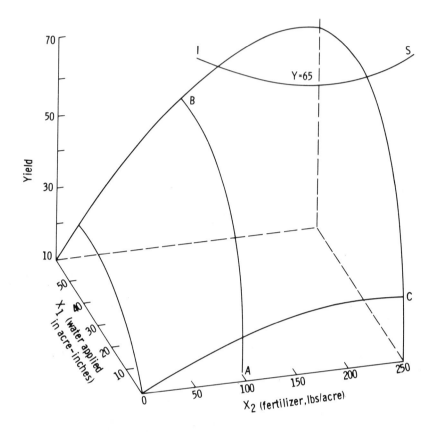

FIG. 2.5. A response surface for water and fertilizer.

combinations of two inputs that will produce a given level of output or yield. Imagine making a horizontal slice through the response surface such as *IS* in Figure 2.5. Corresponding to this slice will be a constant output level, for example, 65 units. Thus *IS* is a product or yield isoquant when yield is at 65. It shows all possible combinations of irrigation water (X_1) and fertilizer (X_2) that will produce a yield of 65 units.

Isoquants are readily derivable from Cobb-Douglas functions. Let the input-output relationship be represented by (2.24). If (2.24) is written in terms of X_1 with Y held constant, as in (2.25), combinations of X_1 and X_2 producing that level of Y can be derived.

$$Y = aX_1^{b_1} X_2^{b_2} = 10\, X_1^{.3} X_2^{.2} \qquad (2.24)$$

$$X_1 = (a^{-1} Y X_2^{-b_2})^{1/b_1} = (.1\, Y X_2^{-.2})^{3.33} \qquad (2.25)$$

For example, 50 units of Y can be produced with 16 acre-inches of water and 49 pounds per acre of fertilizer or with 20 acre-inches and

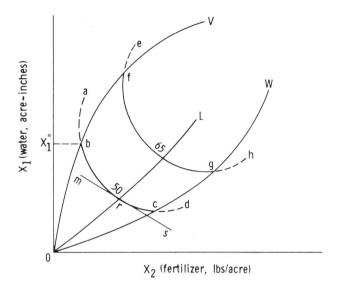

FIG. 2.6. Isoquants (*ad* and *eh*), ridgelines (*OV* and *OW*), and an isocline or expansion path (*OL*) for X_1 and X_2 in production of *Y* at levels of 50 and 65.

35 pounds, respectively. A family of isoquants can be derived by specifying alternative, constant levels of output and solving for the value of X_1 consistent with each value of X_2. Two isoquants are specified in Figure 2.6.

The slope at each point on an isoquant in Figure 2.6 represents the rate at which X_2 must be substituted for X_1 to maintain output at the constant level. The marginal rate of substitution (*MRS*) between X_1 and X_2 is derived by equating the total differential of (2.23) to zero as in (2.26a) and then rearranging terms as in (2.26b). The h_1 and h_2 coefficients represent the respective marginal physical products of X_1 and X_2 in production of Y. According to (2.26b), the *MRS* between X_1 and X_2 equals the inverse of the ratio of their marginal products.

$$dY = h_1 dX_1 + h_2 dX_2 = 0 \qquad (2.26a)$$

$$MRS = -dX_1/dX_2 = h_2/h_1 = MP_2/MP_1 \qquad (2.26b)$$

Equation (2.26b) is the marginal rate of substitution of X_2 for X_1.

Using the Cobb-Douglas function in (2.24) we define the marginal rate of substitution of X_2 for X_1 as the marginal product of X_2 (the partial derivative of Y with respect to X_2) divided by the marginal product of X_1 (the partial derivative of Y with respect to X_1) as shown in (2.27a) below.

$$MRS = MP_2/MP_1 = \frac{\partial Y}{\partial X_2} \bigg/ \frac{\partial Y}{\partial X_1} = 2X_1^{.3} X_2^{-.8}/3X_1^{-.7} X_2^{.2} \qquad (2.27a)$$

Simplifying the last term of (2.27a) we obtain the numerical value of the *MRS* of X_2 for X_1 as the value in (2.27b). As X_2 is increased by successive increments, progressively less of X_1 can be replaced if yield or production level is to remain constant. Hence, the marginal rate of substitution declines, as also is suggested by the isoquants *ad* and *eh* in Figure 2.6 where the slope declines toward the X_2 axis.

$$MRS = .67 X_1 X_2^{-1} \qquad (2.27b)$$

Returning to our earlier example, when X_1 is 20 and X_2 is 35, the *MRS* of X_2 for X_1 is 0.38. If X_2 is increased to 49 and X_1 is 16, the *MRS* declines to 0.22. Of course, if the production function is linear, the marginal products and marginal rates of substitution for resources will be constants.

In Figure 2.6, isoquant *ad* has a negative slope between points *b* and *c*, but a positive slope over the segments *ab* and *cd*. For isoquant *eh*, the slope is negative for the segment *fg* but positive over the segments *ef* and *gh*. Only the negatively sloped portion is an economically relevant range for combining resources such as water and fertilizer. Whereas the resources are substitutes within this range, they are complements over the segments of the isoquants that are positively sloped. Over the complementary range, use of more of one resource also requires use of more of the other to maintain yield or output at a given level. As demonstrated later, this knowledge of marginal rates of substitution and isoquants is important for determining economically efficient levels and combinations of different resources.

Ridgelines and isoclines

Lines *OV* and *OW* in Figure 2.6 separate positive and negative sloped portions of the isoquants (that is, segments of the isoquants defining resource substitutability and complementarity). Lines such as *OW* and *OV* separating substitution and complementary ranges of the isoquants are termed ridgelines. They are isoclines where the marginal rate of substitution of one input for the other is zero. An isocline is a line connecting points of equal slope on adjacent isoquants; that is, they connect points of equal *MRS*s on successive isoquants. The slope of the isoquant at a particular point is the *MRS* at this point. The line *ms* in Figure 2.6 is tangent to isoquant *ad* at point *r*. Hence, the slope of line *ms* defines the marginal rate of substitution of isoquant *ad* at point *r*. An equation for an isocline can be derived by setting, for example, (2.26b) equal to a constant, *k*, then rewriting the equality in terms of one of the inputs as in (2.28). The constant *k* may represent the price ratio of X_2 to X_1, or P_2/P_1.

$$X_1 = k b_1 b_2^{-1} X_2 \qquad (2.28)$$

When *k* is expressed as a price ratio, the combination of X_1 and X_2 corresponding to the intersection of an isocline and an isoquant represents the least-cost mix of X_1 and X_2 for producing the output associated with that isoquant, given the existing price relationship. Assume $P_1 = 1.35$ and $P_2 =$

0.08. With the earlier defined values for b_1 and b_2, the isocline equation in (2.29) is derived from (2.28).

$$X_1 = .0889 X_2 \qquad (2.29)$$

Using these relationships, an isocline OL is superimposed in Figure 2.6. It intersects isoquant ad at point r. Since ad is tangent to ms at point r, the isocline OL thus designates points on all isoquants where the MRS is equal to the slope of ms. If ms also has a slope equal to the input price ratio, it designates the input combination that will minimize the cost of producing the output level represented by the isoquant.

Isoclines are also termed expansion paths indicating the proper mix of inputs as output is expanded by moving to successively higher yield levels. Linear isoclines emanating from the origin indicate that inputs should remain in fixed proportions as output is increased. Curvilinear isoclines or linear ones not passing through the origin indicate inputs that should be, if cost of each output level is minimized, used in varying proportions as the output level changes.

Up to this point, concepts such as production functions and marginal rates of substitution have been discussed individually. In the subsequent sections, these concepts are integrated into a decision-making framework. The initial situation reflects profit maximization when two inputs are used to produce a single output. Later, a multiple input-output situation is introduced.

Economically optimum input use-levels

Determination of the "best" input use-levels for a two-variable function involving water and fertilizer is simply an extension of the procedures previously described for deriving the optimum use-level for a single variable function. A profit function is given in (2.30) where P_y, P_1, and P_2 are the prices received and paid for the output and inputs, respectively, and FC represents fixed costs.

$$\pi = P_y Y - P_1 X_1 - P_2 X_2 - FC \qquad (2.30a)$$

$$\pi = P_y h(X_1, X_2 | X_3, X_4) - P_1 X_1 - P_2 X_2 - FC \qquad (2.30b)$$

Substituting the production function (2.23) for Y in (2.30), with both X_1 and X_2 variable, the profit equation becomes (2.30b). The assumed goal of the producer is profit maximization. Setting the first-order partial derivatives of (2.30b) with respect to X_1 and X_2 equal to zero, a system of two equations is obtained:

$$\partial\pi/\partial X_1 = P_y(\partial Y/\partial X_1) - P_1 = 0, \text{ or } \partial Y/\partial X_1 = P_1/P_y$$
$$\partial\pi/\partial X_2 = P_y(\partial Y/\partial X_2) - P_2 = 0, \text{ or } \partial Y/\partial X_2 = P_2/P_y \qquad (2.31)$$

Solving this system simultaneously, we have the quantity each of X_1 and X_2 that will maximize profits relative to fixed resources. By substituting these values back in the production function, we estimate Y, yield or production level, consistent with profit maximization. If we simply divide the first equation of (2.31) by the second equation and rearrange terms, the following first-order condition for profit maximization is derived:

$$MRS = MP_1/MP_2 = P_1/P_2 \qquad (2.32)$$

When (2.32) is satisfied, X_1 and X_2 are used in the proper, least-cost combination. This condition is also represented in Figure 2.6 by the point of tangency between the iso-cost line ms and isoquant ad where the cost level, C, representing ms is

$$C = P_1 X_1 + P_2 X_2 + FC \qquad (2.33)$$

and

$$X_1 = (C - FC)/P_1 - (P_2/P_1) X_2$$

with X_1 expressed as a function of X_2 and the specified parameters. As the levels of C, P_1, and P_2 vary, so do the slope and position of the corresponding iso-cost lines as well as the X_1 and X_2 levels that satisfy (2.33). The exception would be if C, P_1, and P_2 are all varied in the same direction and in a fixed proportion. In addition to inputs being used in the correct proportion as in (2.32), inputs must be used in the appropriate quantities as suggested in (2.31), so that output is at the profit-maximizing level.

Implicit in the profit equation of (2.30) is the assumption of an unlimited quantity of "working capital" for acquiring X_1 and X_2. The more realistic situation is that capital is limited. In this case, we need to impose an upper limit on capital availability, K, so that $K \geqq P_1 X_1 + P_2 X_2$. In other words, the amount spent on inputs, $P_1 X_1 + P_2 X_2$, cannot exceed capital available, K. Consequently, (2.30) is reformulated as a constrained profit maximization in (2.34) where λ is a Lagrangean multiplier.

$$\pi = P_y Y - P_1 X_1 - P_2 X_2 + \lambda (K - P_1 X_1 - P_2 X_2) - FC \qquad (2.34)$$

Setting the first-order partial derivatives of (2.34) with respect to X_1, X_2, and λ equal to zero, we define the levels of input for X_1 and X_2 that maximize profits to fixed resources. The following set of equations is derived accordingly:

$$\partial \pi / \partial X_1 = P_y \partial Y / \partial X_1 - P_1 - \lambda P_1 = 0$$

$$\partial \pi / \partial X_2 = P_y \partial Y / \partial X_2 - P_2 - \lambda P_2 = 0 \qquad (2.35)$$

$$\partial \pi / \partial \lambda \;\; = K - P_1 X_1 - P_2 X_2 = 0$$

The third equation of the set indicates that the amount of capital spent on X_1 and X_2 cannot exceed the quantity available, K. The conditions in (2.35) must be simultaneously attained if profits are maximized subject to the capital limits. In comparison with the unconstrained conditions in (2.30a) and (2.31), the magnitudes of the marginal products for (2.35) are those in (2.36). Hence, the capital constraint can cause marginal products of the resources to be higher, meaning that for profit maximization a smaller level and different mix of the inputs will be used as compared to the unrestrained case.

$$\partial Y/\partial X_1 = (1 + \lambda) P_1/P_y \text{ and } \partial Y/\partial X_2 = (1 + \lambda) P_2/P_y \qquad (2.36)$$

The parameter λ can be interpreted as the increase in profit, π, corresponding to or resulting from an incremental increase in K. The term λ can also be viewed as the opportunity cost of using the last unit of capital in producing Y rather than in its most remunerative, alternative use. Units of K, for example, could be invested in a savings account or used in another production enterprise. When producers face constraints other than capital availability and have objectives other than profit maximization, (2.34) must be modified accordingly.

Assume $K = 20$ per acre and P_y, P_1, and P_2 are at the previously assumed levels. If the input-output relationships are represented by (2.24), the solution to (2.34) and hence (2.35) is derived as about 8.9 inches of X_1 (water) and about 100 pounds of fertilizer per acre or X_2; λ is estimated at 1.814. That is, if an additional dollar of capital was available for acquiring inputs, profits would be increased by 1.814 dollars. If K were increased to 25, the profit-maximizing levels for X_1 and X_2 would increase to 11.1 and 125, respectively, with λ reduced to around 1.623. As the amount of K is increased, other factors unchanged, the opportunity cost associated with the last increment of working capital is reduced. The second-order conditions for a constrained profit maximization are satisfied when the determinant $D > 0$ where

$$D = \begin{vmatrix} \partial^2 \pi/\partial(X_1)^2 & \partial^2 \pi/\partial X_1 \partial X_2 & -P_1 \\ \partial^2 \pi/\partial X_1 \partial X_2 & \partial^2 \pi/\partial(X_2)^2 & -P_2 \\ -P_1 & -P_2 & 0 \end{vmatrix} \qquad (2.37)$$

If these conditions are not met, profit minimization rather than maximization could occur. When (2.37) is quantified for the above example under profit maximization, $D \cong 0.003$ when 8.9 acre-inches of X_1 and 100 pounds of X_2 per acre are applied to the fixed, unspecified levels of X_3 and X_4.

Product supply function

The quantity of Y supplied to the market is dependent on a number of factors. Production decisions are usually made on the basis of anticipated input and output prices and on expected technical relationships.[2] Both are subject to allowance for price uncertainty and other exogenous forces, such as environmental conditions.

Returning to the earlier two input—one output formulation, the total cost function is written as

$$TC = P_1 X_1 + P_2 X_2 + FC \qquad (2.38a)$$

$$= C(Y) + FC \qquad (2.38b)$$

Total variable cost is rewritten in (2.38b), C, as a function of the output level, Y, and is represented by $C(Y)$. The latter implicitly incorporates the condition that X_1 and X_2 are used in their least-cost combinations when producing any level of Y. The profit equation then becomes

$$\pi = P_y Y - C(Y) - FC \tag{2.39}$$

with $d\pi/dY = P_y - dC(Y)/dY = 0$ and

$$P_y = MR_y = SMC_y \tag{2.40}$$

In (2.40) the price of the commodity, P_y, is equal to the marginal revenue, MR_y, of the commodity under competitive conditions. In turn, if profit is maximized, price or marginal revenue must be equal to the short-run marginal cost, SMC_y, of the product. The short-run marginal cost represents the cost of producing one more unit of the commodity. Profit is maximized under the conditions of (2.40), just as it is under the conditions of (2.31) or (2.35). In other words, differentiating (2.39) with respect to Y, profits are maximized when the marginal revenue generated by an incremental unit of output is equal to the short-run marginal cost of producing that output, as in (2.40). Second-order conditions require that $d^2\pi/dY^2 = -d^2C(Y)/dY^2 < 0$. That is, SMC_y must be increasing at the profit-maximizing output.

Product supply functions are readily derivable from Cobb-Douglas functions. (Computations are somewhat more complicated for other functions.)

As an illustration, let us return to the production function in (2.3). Expressing X_1 as a function of Y, we have (2.41). Now substituting the value of X_1 in (2.41) into (2.38a) when only X_1 is variable, we have the total cost function in (2.42).

$$X_1 = (k^{-1} Y)^{1/b_1} \tag{2.41}$$

$$TC = P_1 (k^{-1} Y)^{1/b_1} + FC \tag{2.42}$$

Taking the first derivative of (2.42) with respect to Y, we obtain (2.43) as the short-run marginal cost, SMC.

$$SMC = b_1^{-1} k^{-1/b_1} P_1 Y^{(1-b_1)/b_1} \tag{2.43}$$

By equating the short-run marginal costs to the price of the product, P_y, we have defined profit-maximizing conditions. Thus if we set SMC in (2.43) equal to P_y and solve for the value of Y, we obtain (2.44), which is

$$Y = (b_1 k^{1/b_1} P_1^{-1} P_y)^{b_1/(1-b_1)} \tag{2.44}$$

a normative supply function in terms of Y. The value of Y increases with larger values of b_1, k, and P_y but declines as input price, P_1, is greater.

We can compute the elasticity of supply with respect to either the price of the commodity or the price of the resource. The elasticity with respect to commodity price (ϵ_y, P_y) is $(dY/dP_y)(P_y/Y)$ while the elasticity with respect to input price (ϵ_1, P_1) is $(dY/dP_1)(P_1/Y)$. Computed for the short-run normative supply function in (2.44), the supply elasticities are given in (2.45) and

(2.46), respectively, with respect to commodity price and input price. As the value of b_1 increases, the price elasticity of supply will be greater.

$$\epsilon_{Y,P_1} = -b_1/(1 - b_1) \tag{2.45}$$

$$\epsilon_{Y,P_y} = b_1/(1 - b_1) \tag{2.46}$$

When both X_1 and X_2 are permitted to vary, as in (2.24), the corresponding product supply function is derived as

$$Y = [a(b_1)^{b_1}(b_2)^{b_2}(P_y P_1^{-1})^{b_1}(P_y P_2^{-1})^{b_2}]^{1/(1-b_1-b_2)} \tag{2.47}$$

and

$$Y = [5.05 \, (P_y P_1^{-1})^{.3} (P_y P_2^{-1})^{.2}]^2$$

After specifying values for the prices in (2.47), the corresponding output levels that maximize profits can be derived.

If the price elasticities for (2.47) where both X_1 and X_2 are variable were derived and compared to those for (2.44), the elasticities associated with X_1 and X_2 in (2.47) would be higher than for the short-run model in (2.44) where X_1 is variable and X_2 is fixed. Hence, producers should have greater flexibility and price responsiveness when more inputs (that is, water, fertilizer, and other inputs, collectively) are variable in a long-run context.

To refer to (2.44) and (2.47) as product supply functions is to implicitly assume that quantities produced and marketed are identical and that the price and technical coefficients are known with certainty. Further, no constraints on resource availability exist and the producer's goal is to maximize profits. Under real world conditions, uncertainty, restraints on rates of change, psychological characteristics of farmers, fixity in resources, and other factors can cause the normative supply derived as above to be greater than actual supply.

DYNAMIC FUNCTIONS

The concepts discussed in earlier portions of this chapter represented decision-making criteria within a static framework. That is, time is not included as a variable. Time is, however, implicit in nearly all production processes. In the earlier discussion of $Y = kX_1^{b_1}$, b_1 represents some "average" effect of application(s) of X_1 over the production period. The relationship could be reformulated as

$$Y_t = kX_{1t}^{b_1}t \tag{2.48}$$

where t ranges from zero to the end of the production period. Recalling that X_1 represents acre-inches of water applied, values for b_{1t} would vary with the irrigation regime and stage of plant growth. If Y_t is defined as plant growth over a particular time period, b_1 for the first irrigations would be expected to be high relative to any irrigation close to harvest. If Y_t is defined as yield, the b_1 corresponding to an irrigation during a "critical period" of pollination and fruiting would be relatively high. The b_{1t} are, however, averages over the t irrigation applications.

Measures of Y_t are required for estimating the corresponding b_{1t}. These are difficult to obtain. Plant growth as represented by height would not necessarily reflect physiological growth. If Y_t is designated as yield, any procedure for imputing final yield among the various stages of plant growth would be ambiguous. If (2.48) were quantified, concepts such as AP and MP would be derivable for each of the t time periods.

As discussed in subsequent chapters, linear and dynamic programming are tools for incorporating some dynamic aspects of production. Through a system of transfer activities, for example, capital flows through time can be built into linear programming models. Changing technical coefficients, such as the b_t in (2.48), can also be included in the analyses.

In Chapter 13, dynamic programming is used to determine optimum water management policies for operating a reservoir. Following estimation of the distribution of expected streamflows and demand for water through a specified planning period, the set of water management policies expected to maximize total farm income over that planning period is derived.

COMPLEXITY IN REAL WORLD SITUATIONS

Most decision-making situations are characterized by competing production enterprises, each requiring several variable and fixed production inputs. Production constraints such as labor and capital availability usually have time dimensions. When the scope of decision making is expanded to the firm-household context, goals other than pure profit maximization are likely important. The decision maker's preferences may be reflected in resource availabilities as well as in enterprises considered. Work-leisure preferences, for example, condition the quantity of family labor available throughout the production year. Hired labor may be profitably used, but the producer may prefer not to have this additional managerial responsibility. If available, off-farm, part-time work provides additional income and perhaps psychic returns. The decision maker's attitude toward risk aversion affects his decision to invest available capital within the production unit or in an off-farm investment.

As the decision-making environment becomes increasingly complex, solutions of optimal input use and output patterns are best derived through techniques such as linear programming and simulation. As with all analyses, however, valid solutions and interpretations are dependent on correct model specification and the accuracy and availability of necessary data. The concepts specified above for relatively simple situations are also applicable for decision making within a more complex environment. The production functions discussed also provide the basic input-output data or technical coefficients necessary for linear programming, simulation, or other decision procedures.

FEATURES OF SELECTED PRODUCTION FUNCTIONS

The importance of production functions in quantifying input-output relationships and in economic decision making was discussed in Chapter 2. A large number of alternative functional forms exist. These have varying applicability for estimating biological relationships. Several of these forms are specified in this chapter. The properties of these alternative forms are discussed and their applicability for abstracting plant-water-soil relationships is assessed. In later chapters, fertilizer is also included in the analyses. Finally, a selected number of water-related production function studies are reviewed along with the features of several agronomic concepts used in these studies.

A PRIORI PLANT-WATER-SOIL RELATIONSHIPS

While these relationships are best discussed by agronomists and soil scientists, some of the concepts most familiar to agricultural economists are introduced here. Yield response to water availability is a function of several factors. Soil texture is important. Apart from the chemical properties of soil, soil texture affects the capacity of the soil to hold soil moisture and the rate at which moisture is available for plant use. Hypothetical yield responses to water on two different soil types are depicted in Figure 3.1.

Assume OW_1 of water is applied with identical irrigation scheduling to both soils. The yield from the sandy loam soil is designated as OY_a; yield from the silty clay loam soil having a higher water-holding capacity is represented by OY_b. If, for example, OY_b were to be produced on Soil A, $W_1 W_2$ additional units of water would need to be applied. The greater water-holding capacity of Soil B is further reflected by yield OY_b^0 when no water is applied. OY_b^0 is obtainable through residual soil moisture from earlier irrigation seasons and (or) natural precipitation. The sandy type soil with a low capacity to retain water for plant use has a lesser capability for sustaining plant growth without irrigation treatments. The significance of this simplified example is that water requirements for a specific crop vary with soil characteristics. Irrigation scheduling and the quantity of water to be applied are further affected by climatic conditions and the crop being grown. Sugar beets have higher water requirements than wheat. Water loss through evaporation and plant transpiration is higher in hot, dry climates than in more temperate zones.

Water-holding capacity of soils

Soil texture, that is, the relative percentage of sand, silt, and clay, is an important factor in affecting the capacity of the soil to hold water and the

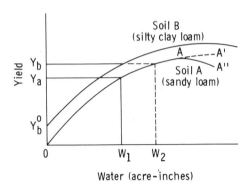

FIG. 3.1. Hypothetical yield response to water applied to a sandy loam and a silty clay loam soil.

force with which moisture adheres to soil particles. Not all this soil moisture is available for plant growth. The range of moisture availability essentially extends from a lower level associated with the permanent wilting point of the plant under consideration to the field capacity level. The permanent wilting point represents "the moisture content of the soil at the time when the leaves of plants growing in that soil first become permanently wilted."[1] The soil moisture content is sufficiently low so that the rate of water absorption by the plant is less than the rate of water lost through plant transpiration. Field capacity, on the other hand, has been defined as "the amount of water a soil will hold against gravity when allowed to drain freely. This point is usually reached in a well-drained soil about 48 hours after the irrigation is completed."[2]   The force with which soil moisture adheres to soil particles is a measure of soil moisture tension. Beringer[3] provides an excellent, nontechnical description of the conditions generating soil moisture tension and the attendant consequences. His comments are paraphrased below. When sufficient water has been applied to raise the soil moisture to, for example, the field capacity level, moisture molecules surround the individual soil particles. Those molecules on the surface of the soil particles are held very tightly while those at some distance are held more loosely. The force at which the latter are held must at least equal the force of gravity; otherwise, this moisture will percolate through the soil beyond the plant root zone. Soil moisture tension under field capacity conditions ranges from about 0.1 to 0.4 atmospheres of tension. As plant roots absorb soil moisture, the most loosely held water molecules are extracted first. As these are removed, soil moisture tension increases and the remaining soil moisture becomes increasingly inaccessible to plant roots. Eventually sufficient moisture is removed so that the soil moisture level drops to the permanent wilting point described earlier. At this point, plant growth ceases, and without additional moisture the plant dies.

Soils having a high clay content also have a greater total soil particle surface area per volume of soil than, for example, a sandy loam soil. The former have a larger capacity for holding soil moisture at any given moisture tension level than the latter. In addition, the range between field capacity and permanent wilting point conditions is wider for clay-type soils than for those having a high sand content. Beringer also cites Parks[4] for pointing out that moisture tension is affected not only by soil texture but by factors such as (a) organic matter, (b) the osmotic effects generated by the level of soluble salts and exchangeable cations, (c) the total pore space and relative distribution that determine the total amount of water that can be held by the soil, and (d) the depth of soil profile together with effective root zone of the plants grown.

A number of these concepts are summarized in Figure 3.2. The soil moisture in the range between field capacity and the permanent wilting point is the moisture readily available for plant growth. As noted earlier, moisture in excess of that capable of being held under field capacity conditions is lost through percolation. This is denoted in Figure 3.2 as gravitational water. Similarly, when moisture tension is at or exceeds the permanent wilting point, the soil moisture adheres so tightly to the soil particles that the moisture is unavailable for plant use.

Reading horizontally in Figure 3.2, when moisture level is at field capacity the water-holding capacity of the clay loam soil is substantially higher than that for the sandy loam. The same is true when soil moisture is at the permanent wilting point. Reading vertically, the shaded areas represent the

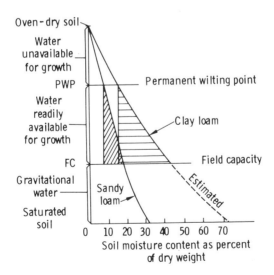

FIG. 3.2. Classification of soil moisture and differences in available and unavailable moisture content of a sandy and a clay soil. (From Kramer, *Plant and Soil Water Relationships*, Fig. 2, 1949, with notational modifications)

water readily available for plant growth. The range of soil moisture between the permanent wilting point and the field capacity is considerably wider for the clay loam than for the sandy loam soil.

The level of soil moisture and plant growth

Beringer[5] and Moore[6] review the alternative hypotheses concerning the availability of soil moisture and its effect on plant growth. One body of thought is based on the contention that plants extract soil moisture with equal facility between field capacity and close to the permanent wilting point. Moore terms this the "equal availability theory."[7] The important implication of this position is that irrigations should be made when available soil moisture is only slightly in excess of that at the wilting point. The hypothesis with most current support is that the rate of plant growth is directly related to the level of soil moisture tension within the active root zone. As noted earlier, this tension of soil moisture stress is basically affected by soil texture, salt content, and the depletion level of available soil moisture. Some of these conditions were summarized earlier in Figure 3.2. Soil moisture tension is further affected by the distribution of roots within the plant root zone, rate of plant transpiration, and time elapsed from the previous irrigation.

Plant growth and subsequent yields are likely susceptible to moisture stress at certain crucial points within the stages of plant growth. Stress at pollination and silking for corn and during flowering for cotton would be expected to have more detrimental effects than stress at points closer to harvest. An example of the general relationship between plant growth (yield) and soil moisture tension is given in Figure 3.3. Shaw and colleagues[8] have conducted

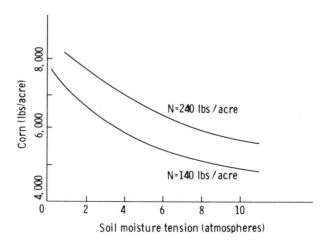

FIG. 3.3.   Relationship between average yield for corn and
soil moisture tension at two levels of fertilization
(Fort Collins, Colorado, 1968).

research in this area. Beringer[9] also cites several studies that support this relationship. The downward sloping curves in Figure 3.3 reflect lower average
yields generated by higher levels of soil moisture tension, which, in turn, reduce the amount of soil moisture available for plant growth. Figure 3.3 is
based on data in Table 6.6. As the level of fertilization increases, the curves
shift upward. The substitution between water and fertilizer can be demonstrated by selecting any yield and then determining the corresponding input
levels. A per acre yield of 6000 is producible with 140 pounds of fertilizer
per acre and relatively higher water applications (relatively low soil moisture
tension) or with 240 pounds of fertilizer per acre and less water (relatively
high soil moisture tension).

Beringer[10] notes two additional factors affecting levels of soil moisture
tension. First, within each soil layer, soil moisture tension increases with
time, t. Second, at any point in time, soil moisture tension decreases with
soil depth. These concepts are summarized in Figure 3.4 where t_0 represents
the initial time period when soil moisture is at the field capacity level and t_n
denotes some unspecified future time period.

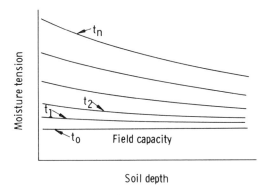

FIG. 3.4. Schematic illustration of the relationship among
soil moisture tension, time, and depth of the
soil profile. (From Beringer, *An Economic
Model*, Fig. 3, 1961)

Moore[11] formulates the relationship between plant growth and soil moisture with slightly different concepts. Three soil types are specified in Figure
3.5. The curve for each would be similar to the curves in Figure 3.2 but inverted. When the percent of available soil moisture depleted equals zero, soil
moisture is at the field capacity level for maximum plant growth. Selecting
any growth rate on the vertical axis, the corresponding percent of available soil
moisture depleted increases as the soils become lighter in texture. Alternatively, when the available soil moisture is depleted to, for example, around 80
percent, a higher growth rate is sustainable on sandy soils than on a clay soil.

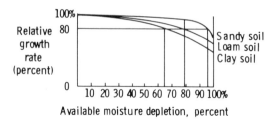

FIG. 3.5.   Variations in how relative growth relates to avail-
            able moisture depletion for sandy, loam, and
            clay soils. (From Moore, *A General Analytical
            Framework*, Fig. 1, 1961)

The available soil moisture is held more tightly by soil particles in the clay
soil, thereby making less available for plant growth.

Irrigation scheduling

Irrigation scheduling relates to the timing of irrigation applications.
Figures 3.4 and 3.5 are useful for developing a hypothetical irrigation
scheduling program. Assume that an irrigation will be made whenever available soil moisture is depleted to, for example, 65 percent. Measuring devices
such as tensionmeters can be used for estimating soil moisture levels. During
initial stages of plant growth, plant transpiration is low and the rate at which
available soil moisture is depleted is also relatively low. As plant growth and
extension of the root system proceed, soil moisture is used more rapidly. At
points near harvest, moisture requirements for plant transpiration taper off
again. Soil moisture is also depleted through evaporation. This rate of evaporation is directly related to temperature, humidity, and wind conditions.

A hypothetical scheduling program for a clay soil is depicted in Figure
3.6. Relative growth is maximized when available soil moisture is at field

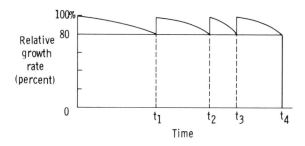

FIG. 3.6.   Hypothetical relationship among rate of plant
            growth, water requirements, and timing of ir-
            rigation for irrigation whenever available soil
            moisture is depleted to 65 percent.

capacity. Just prior to $t_1, \ldots, t_4$, available soil moisture has been depleted to 65 percent. Consequently, an irrigation treatment should be made to return the level of soil moisture to field capacity. The time interval between irrigation treatments is varied to reflect differing moisture requirements at various stages of plant growth. During period $t_0 t_1$, initial plant growth and root development occur; soil moisture is not rapidly depleted during this time. Assume the most rapid growth occurs during period $t_2 t_3$. Since water requirements for plant transpiration are high, the time interval between the irrigation at t_2 and t_3 is relatively short. When an irrigation is made, for example, at t_1, sufficient water is applied to permit maximum growth subject to plant growth in the preceding period.

SPECIFICATION OF PRODUCTION FUNCTION

The parameters of alternative production functions affect the shape of yield response curves, as in Figure 3.1. Considering Soil A, response curve OA' might be derived with the Cobb-Douglas function in (3.1) with k and b as parameters.

$$Y = kX_1^b \qquad (3.1)$$

Assume $b < 1.0$ and k represents the contribution to yield generated by inputs used at fixed levels. If X_1 equals 0, yield is zero. More importantly, since yield continually increases with higher levels of X_1, a maximum yield is not defined. A continually increasing yield, however, is not consistent with most observable biological relationships. Fertilizer, for example, can be applied in such large quantities that plant growth is adversely affected and yield decreases. To generate a yield response of the form OA'', a second-degree polynomial, as in (3.2), would allow estimation of the downward-sloping portion AA'' in Figure 3.1.

$$Y = b_0 + b_1 X_1 - b_2 X_1^2 \qquad (3.2)$$

Three parameters, b_0, b_1, and b_2, are incorporated in (3.2); b_0 represents the combined effect of the fixed inputs. While (3.1) may be appropriate for the range of yield observations being analyzed, for example, up to point A, if AA'' represents a segment of the actual but unknown yield response, a production function such as (3.2) is necessary.

The specification and discussion of properties of several frequently used production functions are covered below. These properties are compared with *a priori* plant-water-soil relationships to evaluate the applicability and appropriateness of alternative production functions. In subsequent discussions, Y = yield in pounds per acre, W = acre-inches of water applied, and N = pounds of fertilizer applied per acre.

Cobb-Douglas function

A two-variable Cobb-Douglas or power function is given in (3.3).

$$Y = aW^b N^c \qquad (3.3)$$

This is of the same form as the model in (2.24). The function was initially fitted to data for U.S. manufacturing industries during 1899–1922[12] but has also been used to estimate biological relationships. Despite the computational ease of estimating parameters for this function, it has properties generally not representative of plant-water-fertilizer relationships. Both W and N are limitational in the sense that if either equals zero, Y also equals zero. As noted earlier, a maximum product is not defined for the Cobb-Douglas. Consequently, a decreasing total product and, in turn, negative marginal products are not possible. Rather the total product curve such as OAA' in Figure 3.1 slopes upward indefinitely.

The concepts of isoquants and isoclines were outlined in Chapter 2. Recall that slopes of specific points on an isoquant represent the marginal rate of substitution between the two inputs under consideration and that isoclines delimit points of equal slope (that is, equal marginal rates of substitution) on successively higher isoquants. The isoquants and isoclines for the Cobb-Douglas function are given in Figure 3.7. At point A, the marginal rates of

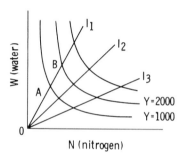

FIG. 3.7. Isoquants and isoclines for a two-variable Cobb-Douglas function.

substitution between W and N are identical to the marginal rates of substitution at point B. That is, the least cost proportions in which W and N are used are invariant with respect to output levels. This property holds because the isoclines are linear and are rays from the origin.

The undefined maximum product, the impossibility of negative marginal products, and invariant marginal rates of substitution associated with each isocline tend to make the Cobb-Douglas function generally less desirable for estimating plant-water-fertilizer relationships.

Mitscherlich-Spillman functions

The pioneering work in quantifying relationships between plant growth and environmental factors was done by E. A. Mitscherlich. In his "law of the physiological relationships," Mitscherlich states that yield could be expanded through increasing levels of any single growth factor so long as that growth factor was not present in sufficient amount to produce the maximum yield.[13] Furthermore, the effects of growth factors on crop yield were considered to be independent.[14]

The exponential form of the Mitscherlich equation for a single variable or growth factor is

$$Y = A\,(1 - e^{-c_1 x_1})\tag{3.4}$$

where Y = observed crop yield; A is the maximum crop yield when x_1 is increased to the limit; x_1 is the growth factor or variable input in the experiment; and c_1 is the constant representing the "effect factor" of x_1 on yield. As x_1 is successively increased, Y becomes asymptotic to A, as in Figure 3.8.

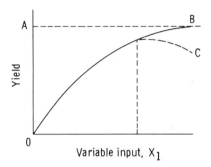

FIG. 3.8. Yield response for a Mitscherlich function for one variable input with others constant and no residual of X_1 in the growth medium.

Briggs[15] notes, however, "In practice it is not the absolute value of x_1 which is measured, but x, the excess of x_1 beyond some value b already present and unknown . . . ," and (3.4) is rewritten as

$$Y = A\,[1 - e^{-c_1\,(x+b)}]\tag{3.5}$$

Baule[16] generalized Mitscherlich's formulation to include several variables, as in (3.6). If any $x_1 = 0$, $Y = 0$. This limitation is relaxed in (3.5).

$$Y = A\,(1 - e^{-c_1 x_1})\,(1 - e^{-c_2 x_2})\,\ldots\,(1 - e^{-c_n x_n})\tag{3.6}$$

Mitscherlich initially postulated the c_1's as constants and equal for all crops, variables, and locations but later partly acknowledged the c_i's may be variable. Hoover[17] reviewed studies where c_i's were found variant. The assumed independence of growth factors as reflected in the constant c_i's has also been criticized and refuted by Balmukand[18] and Briggs.[19] The mathematical interdependence among the c_i's can be demonstrated by the partial derivative of (3.6) with respect to any c_i, for example, c_1. The marginal product of c_1

in (3.7) is dependent on the levels of c_i ($i \neq 1$) and x_i. Any change in c_2, for example, affects $\partial Y/\partial c_1$.

$$\partial Y/\partial c_1 = A\,(x_1 e^{-c_1 x_1})\,(1 - e^{-c_2 x_2})\ldots(1 - e^{-c_n x_n}) \tag{3.7}$$

As noted for (3.4), the yield response curve for a Mitscherlich formulation eventually becomes asymptotic to A. Based on known biological relationships, x_1, for example, fertilizer, can be applied in sufficiently large quantities so that yield decreases, and the MP of x_1 becomes negative. To allow for this possibility, Mitscherlich added "injury factors," denoted as k_i in (3.8), for each x_i. This permits estimation of a yield response curve such as OC in Figure 3.8.

$$Y = A\,[1 - e^{-c_1 x_1}\,e^{-k_1 x_1^2}]\ldots[1 - e^{-c_n x_n}\,e^{-k_n x_n^2}] \tag{3.8}$$

When two variables are introduced such as irrigation water, W, and fertilizer, N, (3.5) is rewritten as (3.9)

$$Y = A\,[1 - e^{-c_1(w+W)}]\,[1 - e^{-c_2(n+N)}] \tag{3.9}$$

where A is the maximum yield when W and N are increased to their limits; c_1, c_2 are the factors representing the efficiency with which the plant uses available water and fertilizer, respectively; w is the available soil moisture when $W = 0$; and n is the residual or preplant fertilizer in the soil when $N = 0$.

Because c_1, c_2, w, and n relate to real phenomena affecting plant growth, the Mitscherlich function has considerable theoretical appeal. If known, these values are substituted into (3.9), and estimates of A, c_1, and c_2 are derived as the function is fitted to the experimental data. If estimates of w and n are not available, (3.9) can be rewritten as (3.10) where $B_1 = e^{-c_1 w}$ and $B_2 = e^{-c_2 n}$.

$$Y = A\,[1 - B_1 e^{-c_1^* W}]\,[1 - B_2 e^{-c_2^* N}] \tag{3.10}$$

Spillman[20] developed an exponential function with some features similar to the Mitscherlich formulations. The general form for the two variable Spillman functions is given in (3.11) where A is the maximum response.

$$Y = A\,(1 - R_w^W)\,(1 - R_n^N) \tag{3.11}$$

The response surface for a one-variable Mitscherlich or Spillman model is similar to curve OB in Figure 3.8. The response surface is asymptotic to A, the maximum yield. The form of the isoquants and isoclines is depicted in Figure 3.9. The isoquants are asymptotic to the W and N axes, indicating that W can never substitute completely for N and vice versa. The isoclines begin at the origin, are curved, and approach linearity. The isoclines do not converge because the response surface is asymptotic to a plane rather than reaching a definite maximum point at A. Because the isoclines are curved, the marginal rates of substitution between W and N change at successively higher output levels, and with invariant price relationships a different mix of W and N is used for each level of output.

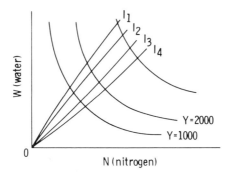

FIG. 3.9. Isoquants and isoclines for a two-variable Spillman function.

The principal limitation of the Spillman function is that R_w and R_n are constants, implying that the marginal products for W bear a constant proportion to each other. The same holds for N. There is little evidence to indicate that this property is applicable to plant-water-fertilizer relationships.

Polynomial forms

Polynomials of varying degrees or order are often used to estimate input-output relationships.[21] The degree is represented by the highest value of the exponents in the model. The polynomial in (3.2), for example, is of degree two. These forms are especially appropriate when the input-output relationship is similar to OC in Figure 3.8 where the marginal product becomes negative and yields decline.

The basic polynomial form is derived from a concept known as a Taylor's series expansion. The principle behind this concept is that "the limit of a sequence (if it exists) can thus be written as the sum to infinity of a convergent infinite series. Any member of the sequence and the sum of any number of terms of the series can then serve as an approximate value of the limit."[22] Assume $Y = f(X)$ with continuous derivatives where

$$f(X) = a_0 + a_1 X + a_2 X^2 + \cdots + a_n X^n \tag{3.12}$$

Through a series of operations described by Allen, (3.12) is written as (3.13) when $X = 0$. The term f' represents the first derivative of $f(X)$ with respect to X, f'' the second derivative, and so on. Assume $X = a$.

$$f(X) = f(0) + f'(0)\frac{X}{1!} + f''(0)\frac{X^2}{2!} + f'''(0)\frac{X^3}{3!} + \ldots \tag{3.13}$$

"It is then possible, subject to the conditions named, to express the value of the function at any point $(a + X)$ in the neighbourhood of $X = a$ as a series in ascending powers of X, the coefficients involving only values at $X = a$."[23] If

X is small, that is, the new value of X is close to a, (3.13) is reformulated as (3.14). Expressing $f(a) = b_0$, $f'(a) = b_1$, and $f''(a)/2! = b_2$, (3.14) is approximated by the quadratic form in (3.2).

$$f(a + X) = f(a) + f'(a)\frac{X}{1!} + f''(a)\frac{X^2}{2!} + f'''(a)\frac{X^3}{3!} + \dots \qquad (3.14)$$

When two variables are included, (3.15) is generated where $f''_{12} = \partial^2 f/(\partial X_1 \partial X_2)$. If X_1 and X_2 are close to a_1 and a_2, respectively, (3.15) can be approximated by the familiar quadratic form in (3.16) where $b_0 = f(a_1, a_2)$ and so on. Equation (3.16) is generalizable to a larger number of variables. If the quadratic variables are omitted from (3.16), a linear function, that is, a polynomial of degree 1 is derived.

$$f(a_1 + X_1, a_2 + X_2) = f(a_1, a_2) + (f'_1 X_1 + f'_2 X_2)_{a_1, a_2}$$
$$+ 1/2! (f''_1 X_1^2 + f''_2 X_2^2 + f''_{12} 2 X_1 X_2)_{a_1, a_2} + \dots \qquad (3.15)$$

Recall that (3.16) represents only a segment of the larger expression for the Taylor's series expansion. Also, (3.16) is based on the assumption that the values for X_1 and X_2 are relatively small. All estimated models are approximations in various degrees.

$$Y = b_0 + b_1 X_1 + b_2 X_2 + b_3 X_1^2 + b_4 X_2^2 + b_5 X_1 X_2 \qquad (3.16)$$

Quadratic (second-order polynomial) function. When (3.16) is fitted to data for Y, X_1, and X_2, a response surface similar to the one depicted in Figure 3.10 is generated. Curve OC is the same form as the yield response curve OA'' in Figure 3.1. The X_1^2 and X_2^2 variables would permit the response surface to curve downward and to exhibit negative marginal products at high use-levels for X_1 and X_2. With the quadratic function, the marginal product curve is linear. This latter property does not appear to be consistent with most agronomic relationships, but it may not be a serious limitation. Finally, the marginal products are not represented by a fixed ratio to each other as in the Spillman function.

An isoquant II' is superimposed on the response surface in Figure 3.10. A family of isoquants and isoclines is drawn in Figure 3.11. Since a maximum output is definable for a quadratic function, the isoquants and isoclines converge at the point of maximum output. The isoclines corresponding to a quadratic function are linear but are not forced through the origin of the input plane. If $k = b_2/b_1$, one isocline will pass through the origin. Isoclines IC_1 and IC_4 also represent ridgelines. To the left of point Q, the marginal product of water is negative and both water and fertilizer must be increased if output is to be maintained. To the right of Q', the marginal product of fertilizer is negative. Since isoclines are also expansion paths, those isoclines not passing through the origin delineate a changing proportion of W and N as they intersect successively higher isoquants or yields at points where the marginal rate of substitution between the two inputs is invariant. This feature seems consistent with known agronomic relationships.

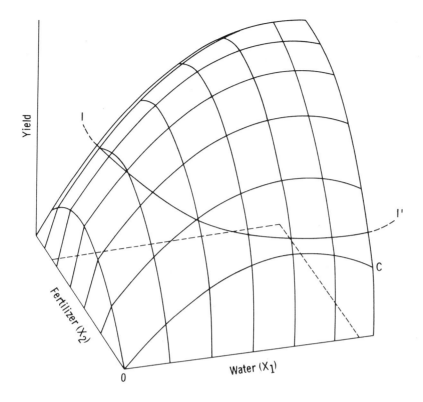

FIG. 3.10. Hypothetical response surface for a two-variable quadratic polynomial.

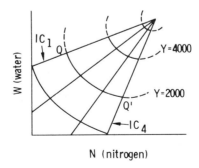

FIG. 3.11. Isoquants and isoclines for a quadratic polynomial.

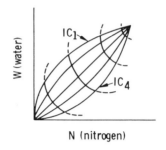

FIG. 3.12. Isoquants and isoclines for a square root
polynomial.

Square root function. Recall that the variables included in (3.16) represent the first few terms from a Taylor's series expansion. Various transfor-
mations can be made on X_1 and X_2. Let $X_1 = (X_1)^{.5}$ and $X_2 = (X_2)^{.5}$.
Equation (3.16) is then rewritten as (3.17), conventionally known as the
square root function. This function has properties similar to those described
for the quadratic. The marginal product curve for either W or N declines at a
decreasing rate while those for the quadratic were linear. This feature of the
square root function seems consistent with agronomic relationships.

$$Y = g(X_1, X_2) = b_0 + b_1 X_1 + b_2 X_2 + b_3 X_1^{.5}$$
$$+ b_4 X_2^{.5} + b_5 X_1 X_2 \qquad (3.17)$$

The isoquants and isoclines for a square root function are given in Figure
3.12. With a positive interaction term in (3.17), the isoclines are curved and
pass through the origin. As with the quadratic form, the substitution rates
between W and N change at higher levels of output as does the proportion in
which W and N are used.

Three-halves or 1.5 polynomial function. Consider an additional transformation of X_1 and X_2 in (3.16). Let $X_1 = X_1^{.75}$ and $X_2 = X_2^{.75}$. Substituting these into (3.16), (3.18) is generated. This formulation, however, is
extremely difficult for estimating isoquants and isoclines. Equation (3.18)$'$
is generally used to approximate the 1.5 polynomial in (3.18). The model in
(3.18)$'$ has several properties similar to the square root function. The marginal product of either W or N, however, declines at an increasing rate.

$$Y = [X_1^{.75}, X_2^{.75}] = b_0 + b_1 X_1^{.75} + b_2 X_2^{.75}$$
$$+ b_3 X_1^{1.5} + b_4 X_2^{1.5} + b_5 X_1 X_2 \qquad (3.18)$$

$$Y = [X_1^{.75}, X_2^{.75}] \cong b_0 + b_1 X_1 + b_2 X_2 + b_3 X_1^{1.5}$$
$$+ b_4 X_2^{1.5} + b_5 X_1 X_2 \qquad (3.18')$$

Simulating production relationships

Simulation is currently being used for a number of problems in agricultural economics as well as other applied fields. Simulation consists of building a model approximating reality that can then be used to investigate the consequences of alternative decisions under varying conditions.

In recent years, several models simulating crop growth have been developed. Soil moisture is generally a primary variable in these models. Most crop-water simulation models are based on the work of Shaw and his associates at Iowa State University.[24] Shaw's simulation model is based on the yield response of corn to varying soil moisture conditions. Shaw demonstrated that his model can be used satisfactorily to explain the reduction in potential yield of corn grain under varying degrees of soil moisture stress at different stages of the growth cycle. The soil moisture–plant growth simulation model is developed to estimate the timing and the amount of irrigation water needed to provide adequate soil moisture for optimum yield and to estimate the yield reductions resulting from alternative soil moisture stress conditions. The three main steps in quantifying the simulation model are to (1) estimate daily values for the factors determining the level of atmospheric demand for moisture by the plant, that is, actual and potential evapotranspiration; (2) estimate the daily supply of moisture to the crop and its distribution within the root zone; and (3) estimate the interaction between the demand for and supply of water on economic yield.

Flinn[25] was one of the first economists who attempted to estimate a water response function through simulation. According to him, his model provides a rational accounting method for estimating daily soil levels and relating the daily soil moisture level to evaporative parameters to obtain an index of plant growth. His simulation results are consistent with actual studies and enable him to estimate the optimal irrigation regime for a given crop under various weather and soil conditions.

Dudley[26] used a soil moisture–plant growth simulation model similar to that employed by Flinn to estimate the values of different stages of crop growth in response to specific irrigation strategies. Dudley's main purpose was to generate data for a stochastic, two-state variable dynamic programming model to be used for determining optimal intraseasonal allocation of irrigation water.

CONSTRAINTS IN ALGEBRAIC FORMS

Selection of an algebraic form directly imposes restraints on the nature of response allowed in the function. Each algebraic form imposes its own unique restraints, although some are much more flexible than others. For example, with the Cobb-Douglas function in (3.3), the elasticity of production, the percentage increase in yield for a 1 percent increase in water (W) or nitrogen (N) is always constant at levels b or c for water and nitrogen, respectively. Its isoclines, $W = c^{-1}bkN$ as in Figure 3.7, are always linear for each value assigned k (a stated marginal rate of substitution between W and N).

They pass through the origin and fan out over space to define no unique yield maximum. The marginal physical product, MP, is equal to the elasticity times the average physical product, AP. Hence, for Equation (3.1) $MP = bAP$.

In the case of the quadratic function in (3.2) the average and marginal product relationships are always linear (but they are not in the Cobb–Douglas, Mitscherlich, and others discussed here) and $MP = AP - b_2 X_1$ if we disregard the constant b_0. Maximum yield is at $X = 0.5\, b_1 b_2^{-1}$ and the curve tends to rise sharply to a maximum and to fall sharply in a mirror reflection of its positive portion. Isoclines are always linear for more than two inputs or variables and do not pass through the origin but do intersect at maximum per acre yield as in Figure 3.11. For the square root function in Equation (3.17), the slope of the function generally will not be as great (that is, the curvature will change more gradually) and the average and marginal product relationships will be nonlinear and $MP = AP - 0.5\, a X_1$ where a is the coefficient of the root term in a "fitted form" such as $Y = a X_1^{0.5} - b X_1$. For two inputs such as in Equation (3.17), the isoclines will be nonlinear, do not intersect the origin, but do intersect at the combination of (say) water and nitrogen, which gives maximum per acre yield. In the Mitscherlich-Spillman function, the MP and AP relationships are not linear and the isoclines fan out over input space indicating that no unique maximum point exists on the response surface but that it serves to define a broad plateau. However, the marginal product of each added input will be a fixed ratio of the MP for the previous input. Hence, for Equation (3.11), the successive marginal products for water are $MP_j = R_w MP_{j-1}$, or the marginal product for the jth input of water is R_w proportion of the $j - 1$ input of this resource.

As pointed out previously, all forms of continuous functions possess some special characteristic such as those presented above for a few of the algebraic forms discussed in this chapter. Because of these reasons and the fact that the form of the function can affect the magnitude of inputs that define an

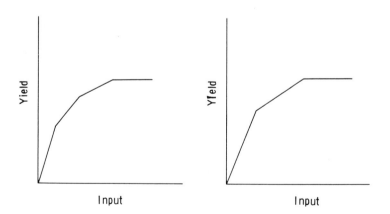

FIG. 3.13. Linear-segmented and linear plateau response models.

optimum in water or fertilizer use, some persons suggest that "form free" functions be used. Others suggest that linear-segmented and linear-plateau functions be used to avoid restraints forced by particular forms of continuous functions. Nelson and Anderson have examined the implications of certain algebraic forms applied to fertilizer data and suggest that where the true form is not known an alternative is to estimate a linear-segmented or linear-plateau model such as that illustrated in Figure 3.13.[27] They present experimental designs for estimating such linear-segmented and linear-plateau responses for a single variable. Somewhat parallel are grafted polynomials as suggested by Fuller.[28] A function approximated as a series of linear segments could provide sufficient detail (and certainly more than currently available) for both improved farm and project planning in better water use. However, difficulties arise when there is a need to estimate water response in interaction with other inputs such as nitrogen. Too, it is unlikely that a yield plateau best describes water and nitrogen response because sufficiently large quantities of either are known to depress yield and result in negative marginal productivity of the input. For these reasons generally, we do not attempt to estimate linear-segmented functions in this study. Linear segments based on the continuous functions are used, however, in certain programming studies that were included in the overall project.[29]

PROCEDURES FOR ESTIMATING PRODUCTION FUNCTIONS

Data on input-output relationships for the plant-water-fertilizer-soil complex are relatively scarce. Several field experiments focusing on a single input, such as different rates of fertilizer application, are available for various crops at various locations. A few secondary sources provide aggregated data of acres planted, production, and a partial accounting of input use-levels. When these measurements are transformed to per acre levels, estimates of input-output relationships are derivable. Such estimates, however, are of little practical use to individual producers, but these averages might be used in studies of regional economic development.

Farm surveys generally provide single-point estimates of certain input-output relationships. That is, producers can indicate an estimate of per acre yield associated with the quantity of seed, water, fertilizer, and perhaps labor applied. These will normally be single estimates per crop in that producers raising corn, for example, tend to apply water and fertilizer at the same rate across the entire field planted to corn. Where land is variable in soil features and (or) topography, producers may adjust accordingly the levels of fertilization and particularly water. When this occurs, the producer may be able to provide two or more estimates of input-output relationships for a crop with each applicable to a specific set of soil and site conditions.

In other cases where producers keep extensive farm records, input-output relationships over time may be estimated. When time is introduced, however, some factors such as crop varieties, fertilizer composition, environmental conditions, and management would be expected to vary. As these factors change, the interyear comparability of the data is decreased. Recall that one stipulation for estimating a production function is that each input is homogenous within the quantity applied. If, for example, fertilizer is applied at a rate of 100 pounds per acre, the entire 100 pounds should have the same composition rather than 50 pounds of 40-0-0 and 50 pounds of 40-30-0. When inputs are not homogenous and (or) intertemporal effects are important, dummy variables to represent these discontinuities can be included in production functions. This technique is discussed later in this chapter.

A production surface could be estimated by aggregating individual producer estimates. Implicit in these production relationships, however, are the same factors mentioned earlier such as differences in soil characteristics, applications of other inputs, and levels of management. This aggregation process involves both aggregation and specification bias, concepts that will be discussed later. A replication of producer estimates in the sense of having several years of data from the same production unit would be preferable in that soil and management inputs would be partially controlled and could be held at essentially constant levels. Replication of observations also permits more reliable statistical analyses of the data as well as more valid interpreta- tions of the data. But even in this case, if learning processes are affected by interyear production experiences, management is not a homogenous input. Also, interyear yields would need to be adjusted for differences in inherent soil fertility and other soil features associated with crop rotation programs.

As noted above, data from farm surveys and from secondary sources usually have several features limiting their usefulness for estimating produc- tion functions. When they are used in production function analyses, these limitations should be specified to permit a more valid interpretation of the results. Field experiments usually provide sufficient yield observations for estimating a continuous input-output relationship. Data generated by field experiments, however, also have limitations. In each experiment, the use- levels of inputs such as seed, water, fertilizer, and perhaps to a lesser extent, management are controlled. They are controlled in the sense that they can be predetermined and (or) changed in the course of the experiment. Experi- ments are, however, always subject to uncontrollable environmental influ- ences. The quality of the data generated is further affected by the accuracy with which both variable and fixed inputs and subsequent yields are measured. If field experiments generate the data necessary for estimating continuous production functions, why aren't more experiments being conducted? The major obstacle is that they are relatively costly.

TRANSFERABILITY OF DATA

An important question is, Are the input-output relationships generated or observed at location A applicable for predicting yields at location B? As noted earlier, data from field experiments tend to be site specific. To the extent that factors such as plant variety, water quality, soil features, and managerial competence differ between sites A and B, observed yields would also be expected to differ. Even if these factors were the same, the scheduling of inputs at B would need to be identical to A if yields generated at B were to approximate those observed at A. For example, the timing and distribution of water and fertilizer applications in B would need to follow the scheduling of treatments used at site A. One means for increasing the transferability of input-output relationships generated in irrigation experiments is to schedule the applications according to levels of soil moisture tension or available soil moisture.

The really important consideration is that sources and reliability of the data be as fully specified as possible. If this information is missing, the ap- plicability of these data to area A or their possible transferability to area B cannot be fully assessed. Lacking perfect knowledge, the validity of this as- sessment process is one of degree. The nature of the experimental design is an important factor conditioning the transferability of input-output data.

EXPERIMENTAL DESIGN

To a considerable extent, the analysis and transferability of any data are conditioned by methods used in obtaining the data. The objective in designing any experiment is to complete a sequence of steps prior to conducting the experiment to help insure obtaining the maximum amount of information relevant to the problem being researched so that valid inferences from the analyses can be made. This applies to both field experiments and to field surveys. The scope and nature of the experimental design are further affected by time and financial constraints. Kempthorne[1] lists the following nine general steps involved in experimentation: (1) statement of the problem; (2) formulation of hypotheses; (3) devising of experimental technique and design; (4) examination of possible outcomes and reference back to the reasons for the inquiry to be sure the experiment provides the required information to an adequate extent; (5) consideration of the possible results from the point of view of the statistical procedures that will be applied to them to ensure that the conditions necessary for these procedures to be valid are satisfied; (6) performance of experiment; (7) application of statistical techniques to the experimental results; (8) drawing conclusions with measures of the reliability of estimates of any quantities that are evaluated, careful consideration being given to the validity of the conclusions for the population of objects or events to which they are to apply; and (9) evaluation of the whole investigation, particularly with other investigations on the same or similar problems. These together with a specification of the constraints facing the researcher interact to help determine the most appropriate experimental design. Steps (7)-(9) are especially important for selecting the design for subsequent experiments.

The discussion will focus on the design of field experiments. Following a preliminary introduction of some factors affecting the choice of design, the properties of selected more commonly used designs will be satisfied. As will become apparent, this discussion is of a summary form. Several references giving in-depth discussions of experimental design are available.[2]

Objective of the field experiment

The research objectives are an outgrowth of steps (1) and (2) outlined earlier. These considerations are fairly straightforward and can best be illustrated with a few examples. In those experiments included in this study, the 2 treatments incorporated in the design were water and fertilizer applications. In nonirrigated areas, a fertilizer treatment might be combined with variable row spacing. Variable plant populations, planting, and harvesting dates are other possible treatments. Irrigation treatments may be scheduled according to soil moisture conditions or stage of plant growth. Fertilizer can be applied prior to planting and (or) as a side-dressing during the growing season.

As the factors or treatments are increased, so are the possible treatment interactions and, in turn, the complexity of the experimental design. If 3 levels of water and fertilizer are considered, 9 possible interactions result. If the treatments are increased to 4 or 5, the potential interactions increase geometrically to 16 and 25, respectively. As will be discussed later, not all treatment combinations need to be included in the experimental design.

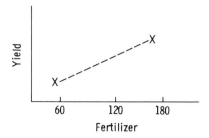

FIG. 4.1. Hypothesized yield-fertilizer relationship when water is fixed at 15 acre-inches and yield response is linear.

The objectives reflect the hypotheses made by researchers concerning the nature of the data to be generated. In agronomic experiments, for example, the range of yield-fertilizer (water) relationships of interest to the researcher affects the size and design of the experiment.

Hypothesis 1. Yield-fertilizer relationships for a given level of water are linear within the range of treatment levels and there is no interaction with water. This hypothesized relation is depicted in Figure 4.1. The yield response curve to fertilizer in Figure 4.1 could be plotted with 2 points or sets of points. In this relatively simple example, the design for 1 block of experimental plots is given in Figure 4.2. A block is a collection of relatively homogenous experimental units (plots). Treatment combinations are usually replicated in the sense that 2 or more experimental units are subjected to the same treatment or treatment combination. The X's in Figure 4.1 usually reflect a group of observed yields rather than single yields. The more variability in observed yields per treatment combination, the greater is the need for repli-

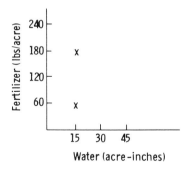

FIG. 4.2. Design for one block of experimental plots to test Hypothesis 1.

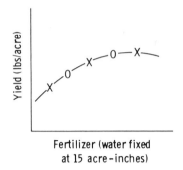

FIG. 4.3. Hypothesized yield-fertilizer relationship when
          water is fixed at 15 acre-inches and yield re-
          sponse is curvilinear.

cation of observations. Adding another block of experimental units and (or)
additional treatment levels would generate more points for estimating the
relationship hypothesized in Figure 4.1.

Hypothesis 2. Yield-fertilizer relationships for a given level of water are
curvilinear and concave to the origin within the range of treatment levels but
have no interaction with water. With nonlinear relationships, as in Figure 4.3,
a larger number of points are necessary for plotting the yield response curve.
The *X*'s on the curve are assumed to represent the more critical points for
plotting. If true, a relatively larger number of observations at these
critical points would permit a more accurate plotting of the yield-fertilizer
relationship. Let *X* represent one replication and 0 designate a single observation. The distribution of treatment combinations for this design is given

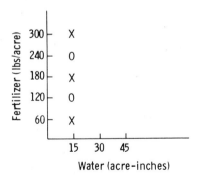

FIG. 4.4. Design for one block of experimental plots to
          test Hypothesis 2.

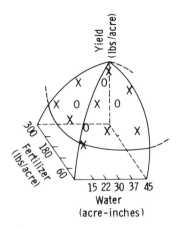

FIG. 4.5. Hypothesized yield-water-fertilizer relationships.

in Figure 4.4. A total of 8 points would be generated for each block of experimental units.

So far, only 1 factor (fertilizer) has been considered. With 2 or more factors, the size of the experiment will likely change and, in turn, the experimental design. Assume Hypothesis 2 is valid for the yield-water-fertilizer relationships in Figure 4.5. A larger number of points, that is, observations from experimental units, are necessary for estimating the response surface in Figure 4.5 as compared with Figures 4.1 and 4.3. If the replication plan in Figure 4.3 were applied to both fertilizer and water, the larger number of points in Figure 4.5 would require more experimental units.

Assume the relationship in Figure 4.3 also applies to yield response curves to both fertilizer and water. Further assume the researcher is interested in 5 treatment levels for fertilizer and 5 for water. If all treatment combinations were included, a total of 25 would exist. Consequently, a minimum of 25 plots per block would be necessary. However, only a single observation per treatment combination would be generated. If all treatment combinations were replicated once, a minimum of 50 plots per block would be required.

If certain points are more important than others, as in Figures 4.3 and 4.5, the layout of the experimental design may be as in Figure 4.6. This design represents an incomplete factorial design. The design is incomplete because not all possible treatment combinations are included. For example, treatment combination water = 15 acre-inches and fertilizer = 120 pounds per acre is not included. The design involves factorials because the fertilizer treatment levels vary over each water treatment and vice versa. The design reflected in Figure 4.6 would require 22 experimental plots per block as compared with 50 if all possible treatment combinations were applied. (An X denotes replication while 0 represents a single observation.) Based on the sequential discussion of Figures 4.1 through 4.6, the size and complexity of

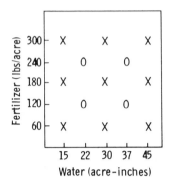

FIG. 4.6. Design for one block of experimental plots to
test Hypothesis 2 with water and fertilizer
treatments.

experimental designs tend to increase as the number of treatments and treatment levels increase.

Nature of the experimental units

The physical features of the experimental units can affect the design and conduct of experiments. The experimental error associated with an experiment can be defined as the failure of two identically treated experimental units to yield identical results. Experimental error has several components including (1) errors of experimentation, (2) errors of observation, (3) errors of measurement, (4) heterogeneity among experimental units and possibly treatment materials, and (5) the aggregative effect of all other factors affecting experimental results but not incorporated in the experimental design.[3]

Blocking. As noted above, one source of experimental error is heterogeneity among the experimental units (plots). Analyses of soil samples taken at several points within the proposed experimental area will give indications for assessing the homogeneity among the plots. Several sources of variation are readily apparent. The soil texture may be highly variable. The topography over the experimental area may be changing. Previous cropping patterns and management practices affect the availability of soil nutrients, organic matter, and soil tilth. Figure 4.7 reflects how cropping patterns and levels of fertilization in period t can affect the homogeneity of experimental units in period $t + 1$. Soil analyses may indicate that the left half of the proposed area for the experiment in period $t + 1$ has a higher level of preplant soil nitrogen than the right half. One means of attempting to identify and control this variation is by blocking.

Blocking involves distributing or allocating the plots to blocks so that the experimental units within blocks are relatively homogenous. That is, blocking is a technique for identifying a source of variation that in the absence of blocking would be embodied in the experimental error. The blocking pro-

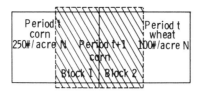

FIG. 4.7. Cropping patterns affecting the need for blocking within the experimental design.

cedure can also be reflected in estimating input-output relationships from
the experimental data. If the treatments in period $t + 1$ are replicated between Blocks 1 and 2, the occurrence of a "block effect" can be estimated
by an analysis of variance. If yields from Block 1 are significantly different
from those in Block 2 at some predetermined tolerance level, a system of
dummy variables can be incorporated into production functions representing
input-output relationships from the experimental area. These dummy variables capture the individual effects that Blocks 1 and 2 have on observed
yields through separating yields into distinct categories by assuming, for example, "dummy" values of 0 and 1 for Blocks 1 and 2, respectively. If the
analysis of variance does not indicate a block effect, experimental conditions
in Blocks 1 and 2 are assumed to be relatively homogenous, and the experimental data can be pooled for analysis. Production functions incorporating
dummy variables are discussed in Chapters 6–10.

Quality of management inputs. One rule of thumb is to keep the experiment as simple as possible but yet obtaining the desired data. Agronomists
with varying degrees of experience and insight would tend to have different
capabilities in determining the level and timing of treatment applications.
Consider the following examples:
    1. An experienced agronomist may have a reasonably good idea of the
levels of carry-over (preplant) soil nutrients from the previous year(s). The
less experienced agronomist (or economist) may need to rely on soil analyses
or may choose to ignore this factor.
    2. An experienced agronomist may have a basis for estimating that
average yields begin declining when, for example, $N = 350$ pounds per acre
and $W = 25$ acre-inches and "average" growing conditions prevail. The less
experienced agronomist (or economist) may need to take a "trial and error"
approach, which may be costly in terms of time and money. If the level of
fertilization is too high relative to that of other inputs (water), the marginal
productivity of fertilizer may turn negative.
    3. In irrigation experiments, the timing of application is important to
plant growth and subsequent yields. Various devices are available for measuring soil moisture and soil moisture tension. An operational knowledge of
these devices such as irrometers, neutron probes, or gypsum blocks is necessary for determining the timing and quantity of water to be applied.
    4. The experienced agronomist is better able to control external sources

of variation. Recall the objective is to examine the effect of specified treatments on yield. The occurrence of disturbing external factors such as
varying plant population, plant disease, insect damage, and careless execution
of treatment applications tends to result in a confounding of these external
effects with the treatment effects.

The researcher's experience and training and that of his assistants likely
have a direct impact on the frequency and magnitude of errors of experimentation, observation, and measurement. These, in turn, affect the level of
experimental error.

Statistical basis for making inferences

Steps can be taken to help insure that the experimental design provides
a basis for making valid inferences from the observed data to the problem
being investigated. Some of the more important procedures are discussed
below.

Randomization. Randomization arranges the field experimental units
into random order thereby converting uncontrolled variation of whatever
source into completely random variation, that is, variation without any
predictable pattern. The implications for estimating production functions
will be discussed later in this chapter.

Randomization helps ensure the absence of any systematic error in, for
example, assigning treatment combinations to experimental units. Contiguous
experimental units would be expected to have more similar features than
those at some distance from each other. Randomization tends to eliminate
this correspondence and remove this component from the estimated experimental error. So far as practicable, different experimental units receiving the
same treatment should be dealt with separately and independently at all
stages at which important errors may arise. One stage is the application of
treatments to the experimental units.

Replication. Repetition of treatment applications permits estimation of
experimental error. The researcher is more interested in average yields per
treatment application than in single observations. That is, replication generates multiple observations and permits determination of average yields.

Replication is also necessary for estimating the "pure error sum of
squares" associated with regression models. This information is necessary for
conducting "lack of fit" test for the models being fitted to the data. Whatever the source of experimental error, replication of the treatments decreases
the error associated with the difference between the average results of two
treatment combinations, assuming the latter were randomly assigned. The
rate at which the experimental error is reduced is predictable from statistical
theory and probability tables.

Factorials. A group of treatment combinations that includes two or more
levels of two or more factors (individual treatments) is termed a factorial

arrangement. Referring to Figure 4.6, the factor $W = 15$ acre-inches is combined with three levels of fertilization. Similarly, the factor $N = 60$ pounds per acre is combined with three levels of water application. As is apparent, the factorial arrangement is incomplete since, for example, the individual water treatments are not combined with all five levels of fertilization. This is sometimes known as fractional replication and is often used when several factors at several levels are included in the experimental design. Cox summarizes the advantages of factorial arrangements as "giving greater precision for estimating overall factor effects, of enabling the interactions between different factors to be explored, and of allowing the range of validity of the conclusions to be extended by the insertion of additional factors."[4]

Range of prediction and projection. Models fitted to experimental data are most valid within the range of treatments included in the experiment. Projecting or extrapolating yields predicted from the model beyond the observed range should be made with caution. Therefore, including several factors covering a relatively wide range of application permits estimation of yield-treatment relationships having a relatively wide range of applicability.

Financial and time constraints

Up to this point, discussion concerning an experimental design appropriate for the problem being researched has not included consideration of financial and time constraints confronting every researcher. The importance of variability among experimental units and quality of management inputs was covered earlier. These are also constraints affecting determination of the proper experimental design.

Field experiments are relatively costly, particularly irrigation studies. Labor and managerial inputs are expensive. Consequently, the amount of money budgeted for the study affects the size of the individual experimental unit, the number of units included in the design, the number of blocks, and the number of treatment levels. Low budget experiments may preclude conducting analyses of soil samples taken throughout the proposed experimental area so that levels of preplant soil fertility can be estimated. Assume that nitrogen is the most limiting soil nutrient and that treatments reflect alternative levels of applications of nitrogenous fertilizer. The information generated by soil analyses is important for selecting a range of treatment levels as well as necessary applications of, for example, phosphorous and potassium to bring soil availability to a normal or nonlimiting level. If the level of preplant soil nitrogen is relatively high, a preliminary crop having high nitrogen requirements may be necessary for reducing available soil nitrogen to a level from which effects of the fertilizer treatments will be observable. Land preparations before and after planting the preliminary crop involve both financial and time considerations.

When irrigation scheduling and treatment levels are based on available soil moisture, devices for monitoring soil moisture levels must be implanted in at least some experimental plots. Labor availability and associated costs would interact in determining whether the devices are placed in each experimental unit or only at representative points throughout the experimental area. Labor and cost constraints would also affect the number of treatments included in the design as well as the number of replications and the inclusion of factorials. The selection of treatment combinations determines, for example, whether the design is a complete or incomplete block.

Conventionally used experimental designs

The properties and requirements for alternative designs are best discussed in texts specifically focusing on experimental design. Only a few of the more commonly used designs will be covered here. These relatively brief descriptions will provide insights into the experimental design most appropriate to the problem under study plus certain constraints confronted by the researchers.

Completely randomized design. This is the basic design. All other randomized designs are variations of this, reflecting constraints on treatments included in the design and the method of assigning treatments to experimental units.[5] In this design the treatments or treatment combinations are randomly assigned to the experimental units within the proposed experimental area. This design should be used only when the experimental units are relatively homogenous. If not, blocking should be used to increase the efficiency of the design. Recall that blocking is a technique for identifying a source of variation that otherwise would be embodied in the experimental error, that is, the unexplainable error.

Consider two factors (treatments). Let W = acre-inches of water applied and N = pounds of fertilizer applied per acre. For simplification, assume that problem is to estimate the linear statistical model in (4.1):

$$Y_{ijk} = \mu + W_i + N_j + (WN)_{ij} + \epsilon_{ijk} \qquad (4.1)$$

where μ is the true mean effect; W_i is the true effect of the ith level of water on yield; N_j is the true effect of the jth level of fertilizer; $(WN)_{ij}$ is the true effect of the interaction of the ith level of water with the jth level of fertilizer; and ϵ_{ijk} is the true effect of the kth experimental unit subjected to the (ij)th treatment combination and $(i = 1, \ldots, I)$, $(j = 1, \ldots, J)$, and $(k = 1, \ldots, K)$. The term ϵ_{ijk} also embodies all other factors such as differences in observation, management, and measurement, which tend to be minimized through randomization. Replications and factorials can also be included.

Assume I and $J = 3$ and $K = 18$. A plot of the experimental design is given in Figure 4.8 where X denotes replication of a treatment combination. The design in Figure 4.8 has a complete factorial arrangement because each of nine possible treatment combinations is included. Since the number of experimental units subjected to each combination are equal, that is, two, this design is also balanced.

While the simple layout of this design has some appeal, two features of the design should be kept in mind. First, the experimental units are assumed to

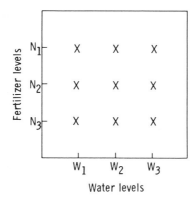

FIG. 4.8. Diagram for a completely randomized design incorporating two treatment variables, each at three levels of application with one replication per treatment combination.

be essentially homogenous. As the number of treatments and replications increase, so does the required experimental area. An increasing amount of heterogenity among experimental units can likely be expected as the experimental area is expanded. Second, since all possible treatment combinations are included, experiments tend to be more costly, especially as the number of treatments and treatment levels increase.

Randomized complete block design. This is similar to the completely randomized design except that the experimental units are grouped into blocks so that the units within each block are relatively homogenous. Variation between blocks is reflected in a block effect. The conditions incorporated in Figure 4.7 may necessitate the need for blocking. Again working with a linear statistical model, (4.1) is modified to include B_m, the true effect of the mth block effect on yield. If two blocks are formed, m in (4.2) becomes $m = 1, 2$.

$$Y_{ijkm} = \mu + B_m + W_i + N_j + (WN)_{ij} + \epsilon_{ijkm} \qquad (4.2)$$

The number of experimental units within each block are equal to the number of treatment combinations of interest. If each treatment application is replicated once, twice as many experimental units are required. Within each block, the treatment combinations are randomly assigned to the experimental units. Figure 4.8 is still appropriate but pertains to only one block of experimental units. If the design incorporates two blocks, Figure 4.8 would be relevant for Block 1 and Block 2.

The advantage of a randomized complete block design is an outgrowth of its necessity. Through blocking, experimental error is reduced when the between-block variation is identified.

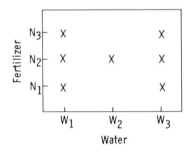

FIG. 4.9.   Diagram for a randomized incomplete block design for one block incorporating two variables.

Randomized incomplete block design. This design is a variation of the
above. The design is incomplete because every possible treatment combination is not included. The other properties are similar to a randomized complete block design. An example of an incomplete block design is given in
Figure 4.9. The model specified in (4.2) also applies to this design. The design may be used when some treatment combinations are of more interest
than others. This phenomenon was depicted earlier in Figures 4.3 and 4.5.
The design in Figure 4.9 is also balanced. In contrast, the design in Figure 4.6
is unbalanced since the number of experimental units subjected to each
treatment combination is unequal. Recall that X and 0 represent a replication
and a single observation, respectively.

The design in Figure 4.9 is less costly than that for Figure 4.8 because
fewer experimental units are incorporated in the former. For the same reason, the increased timeliness of completing operations associated with
Figure 4.9 may further reduce the experimental error.

Other variations

As noted earlier, the most appropriate experimental design can be determined only after a specification of the problem and research objectives, the
level of precision desired, and consideration of time and financial constraints.
Implicit in the above are decisions regarding the number of factors (treatments) to be included in the proposed trial and the means for attempting to
limit the amount of uncontrollable variation, that is, the experimental error.

The addition of a split-plot component to one of the more basic designs
permits addition of another treatment such as variable plant population or
row spacing. Composite rotatable designs permit estimation of second-order
polynomials with a minimum number of treatment combinations. Latin
square and Graeco-Latin squares permit estimation of treatment effects when
a double type of blocking is applied to the experimental units.

The training and experience of the experimenter affect the degree of complexity in experimental design that he can effectively accommodate. A
progression from relatively simple to increasingly complex designs is a normal

sequence. The researcher can build on his accumulating experience. Finally, the validity of statistical analyses and inferences is directly related to the experimental design used. When given a definition of the research problem, the objectives pursued, and the means available, practicing statisticians can usually recommend a design(s) compatible with time, financial, and managerial constraints.

PLOTTING OBSERVED RELATIONSHIPS

Hypotheses concerning the product-yield relationships are implicit in the experimental design. This is evident in selection of the treatments and the levels of application. Refer, for example, to the discussion of Figures 4.1 and 4.3. Following execution of the experiment, analyses of the data generated are compared with the earlier hypothesized relationships. Since experimenters are subject to varying degrees of control, a caution is perhaps appropriate. A researcher may, for example, hypothesize a linear relationship between yield and applications of fertilizer and then choose an experimental design including treatment levels that will generate data to support the hypothesis. Similarly, the impact of varying plant population on yield may be hypothesized as negligible. If the range of plant population treatments is narrow and (or) selected levels of other treatments offset differences in plant population, data may be generated to support the hypothesis. The careful analyst will avoid these potentially tautological exercises; the careful observer will interpret the experimental data and their analyses in connection with the experimental design used. These simplified examples suggest the need for drawing on experience and previous research in accurately defining the problem to be studied. The experimental design should be appropriate to the problem defined.

The experiments in this study relate to plant-water-fertilizer relationships. Sufficient knowledge about biological relationships is known to postulate that yield declines if water or fertilizer is applied in sufficiently larger quantities. In the extremes, the plant dies. Consequently, a curvilinear yield response, as represented in Figure 4.3, may be hypothesized for water and (or) fertilizer with treatment levels selected accordingly. But since many environmental conditions cannot be regulated, field experiments are not fully controlled. Weather conditions at critical stages of plant growth, incidence of plant diseases, and errors of execution may individually or cooperatively cause generation of data inconsistent with hypothesized relationships.

Plots of the experimental data generated provide insights into the nature of the estimated product-treatment relationships. With two treatments, the yield response surface is three-dimensional. Since two-dimensional surfaces are easier to interpret, consider Figure 4.10.

Assume the points in Figure 4.10 represent observed yields associated with N_1-N_6 levels of fertilizer when water is fixed at 20 and 30 acre-inches, respectively. If the fertilizer treatments were confined to the N_1-N_3 range, production would be occurring in the accelerating stage of the production function.[6] If a quadratic formulation, such as in (3.16), were

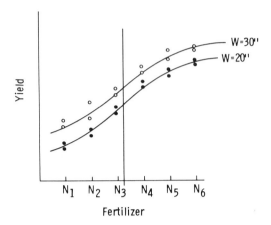

FIG. 4.10. Nature of yield-fertilizer relationship affected
by level of fertilization and water levels.

fitted to these data over the N_1-N_3 range, the signs for the linear and quadratic terms likely would both be positive or estimated coefficients for the
W^2 and N^2 variables might not be statistically significant at conventional
levels. The analyst may be disturbed by this outcome contrary to the hypothesized form, but an examination of the graph of observed yields helps provide
a basis for explaining the properties of the quantified formulation. If the experiment were to be repeated, the level of fertilization should be increased to,
for example, the N_4-N_6 rates. The quantity of water applied could also be
reduced. If the yield response curves in Figure 4.10 were linear within the
range of treatment levels, then a linear function would be most appropriate
for representing the data generated. If the fertilizer treatments ranged from
N_1 to N_6, a cubic function should provide a good fit to the observed data.

Choosing the most appropriate polynomial form for quantifying yield-
water-fertilizer relationships is to some extent a trial-and-error process.
Examining a scatter diagram of the observed relationships may shorten this
process.

ESTIMATING PRODUCTION FUNCTIONS

Following generation of the field experiment data, the next step is to
estimate the yield-treatment relationships. Plotting these relationships
provides some insights. To estimate such concepts as predicted yields and
marginal rates of substitution between inputs, however, the input-output relationships must be quantified. Since subsequent analyses focus on yield-water-
fertilizer relationships, let Y be the per acre yield, W the acre-inches of water
applied, and N the pounds of fertilizer applied per acre. This relationship is
expressed in implicit form as

$$Y = f(W, N) \tag{4.3}$$

where f is an unspecified function form. One of the simplest forms of f is the linear model in (4.4)

$$Y = \beta_1 W + \beta_2 N + \beta_3 WN \tag{4.4}$$

where an interaction term has been appended. Values for Y, W, and N are known. The β_i's are parameters quantifying the relationships implicit in (4.3). If the true relation in (4.3) is linear, the model is fully specified, and there are no errors of measurement, a set of values exists for the β_i so that (4.4) holds. These ifs, however, are rarely satisfied. Consequently, a series of error terms must generally be added to (4.4).

One source of error can be identified as specification bias. This bias occurs when the mathematical model in (4.4), for example, does not reflect the true but unknown relationships embodied in the data. Instead of a linear form, the curvilinear response in Figure 4.5 may be more reflective of yield-water-fertilizer relationships. Consequently, specification bias arises from an incorrect statement of the model and (or) omission of relevant variables. The latter results when the analyst is unaware of these variables or they are not measured or measurable. Let this misspecification be represented by E_1. Even in highly controlled experiments, human errors in execution and measurement are additional sources of variation. Finally, if homogenous experimental units in field experiments were subjected to identical treatments and invariant management, some unexplained variation would still be expected to occur. This variation may be related to specification bias. Let these other sources of error be denoted by E_2. Equation (4.4) can then be rewritten as (4.5).

$$Y = \beta_1 W + \beta_2 N + \beta_3 WN + E_1 + E_2 \tag{4.5}$$

The terms ϵ_1 and ϵ_2 represent deviations of E_1 and E_2 from their respective means. Consequently, (4.6) is an identity.

$$E_1 = \overline{E}_1 + \epsilon_1 \text{ and } E_2 = \overline{E}_2 + \epsilon_2 \tag{4.6}$$

Equation (4.5) can be restated as (4.7) and (4.8) where $\beta_0 = \overline{E}_1 + \overline{E}_2$ and $\epsilon = \epsilon_1 + \epsilon_2$.

$$Y = \beta_1 W + \beta_2 N + \beta_3 WN + (\overline{E}_1 + \epsilon_1) + (\overline{E}_2 + \epsilon_2) \tag{4.7}$$

$$Y_i = \beta_0 + \beta_1 W_i + \beta_2 N_i + \beta_3 W_i N_i + \epsilon_i \tag{4.8}$$

Equation (4.8) represents the general linear model in estimable form. Subscripts have been added to identify individual yield observations. The parameters to be estimated are the β_i's where $(i = 0, \ldots, 3)$. Again, the ϵ_i's represent the unobservable, composite error terms.

In estimating (4.8), the analyst attempts to derive best linear unbiased estimates of the parameters. Best implies that the estimates of β_i, the $\hat{\beta}_i$, have minimum variances. Corresponding to each set of $\hat{\beta}_i$'s is a vector of residuals $[e_i]$ where the e_i's are estimates of ϵ_i in (4.8).

Based upon the definition of ϵ_i in (4.8) and the random assigning of treatments in field experiments, the ϵ_i's are expected to have a zero mean with an unknown variance, σ^2. Further, the ϵ_i's are assumed to be uncorrelated with W_i and N_i, the independent variables in this model. Finally, the ϵ_i's, $i \neq j$, are assumed to be uncorrelated. When these assumptions are satisfied, $\hat{\beta}_i$ can be estimated by ordinary least squares techniques. Since the ϵ_i's in (4.8) have an expected zero mean, the terms in (4.8) are transposed and squared as

$$Q = \epsilon_i^2 = (Y_i - \beta_0 + \beta_1 W_i + \beta_2 N_i + \beta_3 W_i N_i)^2 \qquad (4.9)$$

Ordinary least squares are then used to estimate the β_i, which minimize the error squared terms in (4.9). If the assumptions concerning the error terms are not satisfied, other estimating techniques are required. If the ϵ_i's are correlated with W_i or N_i, for example, the β_i can be estimated by solving a set of simultaneous equations. If the ϵ_i's are correlated, generalized least squares may be used.

Ordinary least squares are based on the Gauss-Markov theorem which states that when the estimated equation is the true equation the estimated parameters are both unbiased and of minimum variance. To the extent the above assumptions are not fully satisfied, the estimating procedure is not strictly correct. The analyst should strive to satisfy these assumptions and to point out those deviations of which he is aware.

The degree of specification bias may be reduced by examining plots of the yield-treatment relationships. If a curvilinear relationship is observed among the experimental data, a second-order or 1.5 polynomial, such as in (4.10) (or other appropriate nonlinear form), should be used rather than the linear form. The 1.5 polynomial in (4.10) can be linearized by transforming W_i and N_i to $W_i^* = (W_i)^{1.5}$ and $N_i^* = (N_i)^{1.5}$.

$$Y_i = \beta_0 + \beta_1 W_i + \beta_2 N_i + \beta_3 W_i^{1.5} + \beta_4 N_i^{1.5} + \beta_5 W_i N_i \qquad (4.10)$$

Models such as (4.10) are identified as being intrinsically linear since they can be transformed into linear forms. When the model is rewritten as (4.11), the ordinary least squares procedure can be used for quantifying the model.

$$Y_i = \beta_0 + \beta_1 W_i + \beta_2 N_i + \beta_3 W_i^* + \beta_4 N_i^*$$
$$+ \beta_5 W_i N_i + \epsilon_i \qquad (4.11)$$

The quantified version of (4.10) estimated for the 1971 corn experiment at Mesa, Arizona, is given in (4.12).

$$Y = 2923.2682 - 295.6678W + 30.4974^{***}N + 46.8803W^{1.5}$$
$$\quad (5985.78) \qquad (571.99) \qquad (9.940) \qquad (67.72)$$
$$- 2.0224^{***}N^{1.5} + .2958WN \qquad (4.12)$$
$$\quad (.351) \qquad (.268)$$
$$R^2 = .818 \quad F = 34.09^{***} \quad LOFF = 2.872^{**}$$

Based on an analysis of variance, there was no "block effect" for this experiment, and data from Blocks 1 and 2 were pooled for analysis. The standard errors of the estimated coefficients are given in parentheses. Determination of the standard errors and statistical significance of the estimated coefficients is discussed at a later point in this chapter. Levels of significance for estimated statistics are *** = 0.01 probability, ** = 0.05 probability, and * = 0.10 probability. Consider first the estimated coefficient for the intercept where $\beta_0 = 2923.2682$. If $W = N = 0$, β_0 can be interpreted as the average effect on yield associated with all excluded variables relevant to the true but unknown model appropriate for the sample of experimental data being analyzed. Recall the derivation of β_0 from Equations (4.5) to (4.9). According to Rao and Miller:

An econometric model (regression equation) is used to explain the behavior of a subpopulation that contains at least one nonzero independent variable. When all the independent variables are zero then the observation does not belong to the subpopulation under investigation, and the regression equation has no valid interpretation.[7]

The R^2, the coefficient of determination, is estimated at 0.818. The equation for estimating R^2 is (4.13) where SS denotes sum of squares (ΣY_i^2) and corrected SS is adjusted for the mean, $\overline{Y}$.

$$R^2 = \frac{\displaystyle\sum_{i=1}^{n}(Y_i - \overline{Y})^2 - \sum_{i=1}^{n}(Y_i - \hat{Y}_i)^2}{\displaystyle\sum_{i=1}^{n}(Y_i - \overline{Y})^2} \qquad (4.13)$$

or

$$\frac{(\text{Corrected } SS) - (\text{Residual } SS)}{\text{Corrected } SS} = \frac{SS \text{ due to regression}}{\text{Corrected } SS}$$

From earlier discussions, the residual SS corresponds to $\Sigma \epsilon_i^2 \cong \Sigma \hat{\epsilon}_i^2$. The R^2 is an indicator of the amount of variation in the observed Y_i that is "explained" by the independent variables incorporated in the model in (4.10). But the R^2 never decreases when additional variables are included. This is true regardless of their relevance toward "explaining" the variance among the observed Y. Equation (4.13) can be rewritten as (4.14) where the expression for s_e^2 is in (4.15).

$$R^2 \cong 1 - (s_e^2/s_Y^2) \qquad (4.14)$$

$$s_e^2 = (n - k)\sigma_\epsilon^2 \qquad (4.15)$$

The number of variables included in the model is $k - 1$. The value of s_Y^2 is unaffected by the number of independent variables. Because of the nature of ordinary least squares, the unobservable level of σ_ϵ^2 does not increase with additional independent variables. However, variables can be added to the point where $n = k$, $s_e^2 = 0$, and $R^2 \cong 1.0$. Summarizing, (4.15) decreases and (4.14) increases as the number of independent variables in the statistical model increase regardless of their relevance in "explaining" variation in Y. Consequently, the estimated level of R^2 may be an inadequate and (or) misleading indicator of appropriateness of the model being estimated. The "F" statistic in (4.12) represents the ratio between the sums of squares due to regression and the sums of squares "unexplained" by the regression model with each divided by its respective degrees of freedom.

Additional criteria for selecting the best model are needed. If the model is expected to both predict and explain, independent variables in the model should reflect *a priori* relationships or previous knowledge of the nature of the data being analyzed. There is little relevance in including, for example, a variable representing the Dow-Jones industrial average in a model for estimating yield-water-fertilizer relationships. Another criterion is a "lack of fit" F test, denoted by the acronym *LOFF*. Under certain conditions, the residual *SS* in (4.13) can be disaggregated into "pure error *SS*" and "lack of fit *SS*."[8] The "pure error *SS*" is estimable if the experimental design included replication of the treatment applications. That is, the "pure error *SS*" and the experimental error are measuring the same phenomena. Using this information, an F test in (4.16) with $(t - k)$ and $(n - t)$ degrees of freedom can be made.

$$LOFF = \frac{\text{Lack of fit } SS/(t - k)}{\text{Pure error } SS/(n - t)} \qquad (4.16)$$

The terms t, k, and n represent the number of alternative treatment combinations, the number of parameters in the model, and the number of observations for the dependent variable, respectively. If the estimated F in (4.16) is statistically significant at a predetermined tolerance level, for example, 5 percent, the estimated model is inadequate for the data being analyzed. A problem in specification may exist. The assumptions implicit in the use of ordinary least squares, which were specified earlier, may have been violated. Randomization in assigning treatment combinations to experimental units helps ensure that the errors, ϵ_i, are uncorrelated. The assumption that the ϵ_i's are uncorrelated with the independent variables can be at least partially evaluated by considering the $(Y - \hat{Y})$ results as estimates of ϵ_i and plotting these against the values of W_i and N_i. Consider the hypothetical distribution of residuals plotted against water treatment levels in Figure 4.11. If the residuals and treatment levels are uncorrelated, the residuals should be approximately evenly distributed about a zero line. Except for one observation, the residuals associated with $W_1 - W_4$ are reasonably well distributed.

The deviant observation, often termed an outliner, may have resulted from errors in measurement in the field or from an error in transcription or tabulation. If the transcription, tabulation, and conversions appear correct, little basis exists for deleting that observation from the analysis, especially since the other three residuals corresponding to W_4 are well distributed. The residuals associated with W_5 present another problem. All four are substantially below the zero line. The person conducting the experiment may be

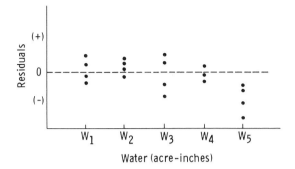

FIG. 4.11. Hypothetical distribution of $(Y - \hat{Y})$ residuals
plotted against water treatment levels.

able to provide some insights or suppositions. If the four observations are omitted from the analysis, this deletion should be fully specified along with the reasons for doing so. Without this specification, the reader will not be able to properly evaluate the analysis.

Returning to (4.12), the residuals generated by this model were plotted against W and N. When the residuals were plotted against the treatment levels of N, they were evenly distributed about a zero line. In plotting the residuals against W, however, the residuals associated with $W = 16.0$, 29.7, and 32.6 acre-inches were not well distributed. The $W = 16.0$ and 29.7 treatments were combined with $N = 0$. (See Table 6.13.) If estimates of preplant soil nitrogen were available and included in the analysis, the distribution of these residuals might be improved. The $W = 32.6$ treatment was not replicated in each of two blocks so only two-yield observations are available. The fewer the number of observations per treatment combination, the more difficult is an evaluation of the plots of residuals. When the yield observations corresponding to these water treatments are deleted, (4.12) is reestimated as (4.17).

$$Y = -17456.2964^{**} + 1071.8934W + 41.5719^{***}N - 84.6556W^{1.5}$$
$$\qquad (8503.90) \qquad (700.72) \qquad (10.11) \qquad (78.41)$$

$$-1.7044^{***}N^{1.5} - .1732WN \qquad\qquad (4.17)$$
$$\quad (.394) \qquad\quad (.283)$$

$$R^2 = .840 \quad F = 29.46^{***} \quad LOFF = .847$$

The $LOFF$ statistic is reduced from 2.872 to 0.847 where the latter is not statistically significant at the 5 percent level. The other important differences between (4.12) and (4.17) are that the estimated coefficient for the intercept in (4.17) is significant at the 5 percent level and the signs for the W and $W^{1.5}$ coefficients are consistent with hypothesized values.

Referring to (4.12), the standard errors corresponding to the estimated regression coefficients are given in parentheses. Since (4.12) is multivariate, matrix notation is most convenient for specifying the formulation for estimating standard errors. A few rows of the X matrix compiled from Table 6.6 for the 1971 corn experiment at Mesa follow, where W = water applied plus preplant irrigation of 8 inches.

$$
X = \begin{bmatrix}
1 & 24.0 & 0 & 117.5 & 0 & 0 \\
1 & 24.0 & 0 & 117.5 & 0 & 0 \\
\cdot & \cdot & \cdot & \cdot & \cdot & \cdot \\
\cdot & \cdot & \cdot & \cdot & \cdot & \cdot \\
\cdot & \cdot & \cdot & \cdot & \cdot & \cdot \\
1 & 34.0 & 0 & 198.3 & 0 & 0 \\
\cdot & \cdot & \cdot & \cdot & \cdot & \cdot \\
\cdot & \cdot & \cdot & \cdot & \cdot & \cdot \\
1 & 30.8 & 85 & 170.9 & 783 & 2618 \\
\cdot & \cdot & \cdot & \cdot & \cdot & \cdot \\
\cdot & \cdot & \cdot & \cdot & \cdot & \cdot \\
\cdot & \cdot & \cdot & \cdot & \cdot & \cdot \\
1 & 49.0 & 340 & 343.0 & 6269 & 16660
\end{bmatrix}
\quad \text{where } Y = \begin{bmatrix}
1144 \\
2095 \\
\cdot \\
\cdot \\
1603 \\
\cdot \\
\cdot \\
2470 \\
\cdot \\
\cdot \\
6753
\end{bmatrix}
\tag{4.18}
$$

Columns of X: $W \quad N \quad W^{1.5} \quad N^{1.5} \quad WN$

The X matrix for this experiment has 44 rows corresponding to the 44 yield observations and 6 columns representing a column of 1's necessary for estimating b_0 and values of the independent variables corresponding to the observed yields. Using matrix notation, the regression coefficients in (4.12) were derived from the following system:

$$
\begin{bmatrix} b_0 \\ b_1 \\ \cdot \\ \cdot \\ b_5 \end{bmatrix} = \begin{bmatrix} 2923.27 \\ -295.67 \\ \cdot \\ \cdot \\ .2958 \end{bmatrix} = (X'X)^{-1}X'Y
\tag{4.19}
$$

where X' represents the transpose of the matrix in (4.18) and $(X'X)^{-1}$ is the inverse of the $(X'X)$ matrix. Y is a vector of dimension 44×1 containing the 44 yield observations. The variance-covariance matrix for the vector of regression coefficients is represented by (4.20) and the estimated form by (4.21).

$$
V(\hat{b}_i) = (X'X)^{-1}\sigma^2 = c_{ii}\sigma^2
\tag{4.20}
$$

$$
\hat{V}(\hat{b}_i) = (X'X)^{-1}s_E^2 = c_{ii}s_E^2
\tag{4.21}
$$

$$\hat{\sigma}^2 = s_E^2 = \sum_{i=1}^{n} (Y_i - \hat{Y}_i)^2/(m - 6) \tag{4.22}$$

In (4.21) and (4.22), estimated parameters are denoted by the overhead The term c_{ii} represents the element at the intersection of the ith row and ith column on the diagonal of the $(X'X)^{-1}$ matrix. The standard errors in (4.12) are derived by quantifying (4.21) and then taking their respective square roots.

One other test statistic is the "t" test for statistical significance of each estimated regression coefficient. The expression for the test statistic is in (4.23) where $b_i^* = 0$. That is, (4.23) tests whether each estimated coefficient is significantly different from zero at some predetermined tolerance level α. The hypothesis is rejected if $t \leqq - t_{(1-\alpha/2,m-6)}$ or $t \geqq t_{(1-\alpha/2,m-6)}$.

$$t = (\hat{b}_i - b_i^*)/s_{b_i} \ (i = 0, \ldots, 5) \tag{4.23}$$

Assume $\alpha = 0.05$ and recall that $n = 44$. Based on a culmulative t-distribution table, the hypothesis is rejected if the t value in (4.23) is ≥ 2.0282 or ≤ -2.0282. If $\alpha = 0.10$, that is, a higher tolerance level is allowed, the hypothesis is rejected if $t \geq 1.6880$ or ≤ -1.6880. Consider the "t" tests for b_0 and b_2 in (4.24) and (4.25), respectively. The estimated coefficient for b_0 is not statistically different from zero at the $\alpha = 0.05$ level. The estimated t-value for b_2, however, exceeds the tabled value of 2.0282, and the hypotheses that $\hat{b}_2 = 0$ is rejected.

$$t = (2923.2682 - 0)/5985.73 = .488 \tag{4.24}$$

$$t = (30.4974 - 0)/9.940 = 3.068 \tag{4.25}$$

Another test statistic that will be used in Chapter 10 is estimation of confidence intervals for the predicted yields, $\hat{Y}$. The variance of $\hat{Y}_0$ for a specified set of values for the independent variables, X_0, is estimated with (4.26). For example, using the first row of data in the X matrix in (4.18), (4.26) can be rewritten as (4.27). $\hat{Y}$ is estimated by substituting $W = 24.0$ and $N = 0$ into (4.12). Again assuming $\alpha = 0.05$, the formulations for estimating the 95 percent confidence interval are given in (4.28) and (4.29).

$$V(\hat{Y}_0) = X_0'(X'X)^{-1} X_0 \sigma^2 \tag{4.26}$$

$$\hat{V}(1335.67) = [1 \ 24 \ 0 \ 117.5 \ 0 \ 0] \ (X'X)^{-1} \begin{bmatrix} 1 \\ 24 \\ 0 \\ 117.5 \\ 0 \\ 0 \end{bmatrix} s_E^2 \tag{4.27}$$

$$\hat{Y}_0 \pm t_{(1-\alpha/2, m-6)} \, s_E (X_0'(X'X)^{-1} X_0)^{1/2} \qquad (4.28)$$

$$1335.67 \pm 2.0282 \, s_E (X_0'(X'X)^{-1} X_0)^{1/2} \qquad (4.29)$$

Multicollinearity

Another phenomenon that can occur in econometric analyses is multi-collinearity among the independent variables. This problem occurs when at least two of the independent variables are so highly correlated that their individual effects on the dependent variable are not discernible. If $Y = F(X_1, X_2, X_3)$ and X_2 and X_3 are highly correlated, $\partial Y/\partial X_2$ or $\partial Y/\partial X_3$ is not meaningful since the separate effect cannot be measured. Estimated coefficients for collinear variates tend to have high standard errors and are, therefore, imprecise. Determination of the correlation among values of the independent variables helps alert the analyst to the possible occurrence of multicollinearity. When X_2 and X_3 are highly correlated, only one may be selected for inclusion in the model. The choice may be on a priori basis and (or) the level of correlation with X_1.

Problems with multicollinearity are more frequent when data from secondary sources are used. When primary data are being generated with field experiments, for example, the experimental design and treatment levels can be chosen so that the problem is avoided. Variables such as W and $W^{1.5}$ in (4.17) are highly correlated. But in predicting yields or estimating the marginal product of water, the relative effect due to the linear and 1.5 terms for water is usually of little interest. Consequently, if models are primarily used to predict rather than explain, multicollinearity is less of a problem, especially if the interrelationships are expected to continue in the future.[9]

Range for projecting yields

Since functions are fitted to a given set of yield and treatment observations, the function selected is only appropriate for the given range of observations. The importance of this limitation is demonstrated by the hypothetical distribution of yields given in Figure 4.12. The curvilinear function designated as AB fits well for the yields corresponding to W_1 through W_5 irrigation levels. If the yield associated with W_6 were to be predicted on the basis of function AB and if actual yields were as plotted, the predicted yield would exceed the actual yield. In this situation, a functional form represented by ABC is required even though AB is suitable for yield-water relationships through W_5. Consequently, any function fitted is most reliable and most appropriate for the range of observations known to the analyst. The estimated coefficients for the response surface may not correspond variable by variable to the true structural representation, but taken on the whole, the fitted coefficients have effects that approximate those of the true but unknown parameters within the range of available observations.

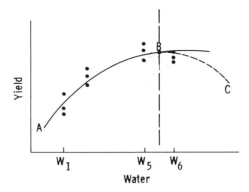

FIG. 4.12. Predicting yields beyond range of observations.

Nonlinear regression

Not all models are linear or intrinsically linear in their parameters. Models of the Mitscherlich form in Chapter 3 are nonlinear. In these situations, ordinary least squares cannot be used to estimate the parameter coefficients, but rather iterative procedures are available for quantifying the models. Three more commonly used procedures are Hartley's linearization,[10] steepest descent,[11] and Marquardt's compromise.[12] Hartley's procedure is used to estimate parameters of Mitscherlich models in subsequent chapters. Because a description of the procedure involves extensive notation, it will not be presented here. Hartley's procedure is iterative in that the values initially postulated for the parameters being estimated eventually converge to estimates of the true parameters. That is, these initial values are successively improved through iterative processes which work toward minimizing the residual sums of squares.

ORIGIN AND FEATURES OF THE RESEARCH PROJECT

The science of economics focuses on the efficient use of resources toward attaining some goal. That goal may be to maximize profits from a farming operation subject to constraints on water and capital availability. At a different level of decision making, the goal may be to allocate water from a reservoir to users through time, for example, to maximize total farm income or to minimize variability of farm incomes. These water management decisions define the constraints on water availability confronted by individual users.

Although water is only one of several inputs in agricultural production, it is a limiting resource in many arid and semiarid areas. As the demand for water in agricultural, industrial, municipal, and recreational uses continues to grow, the efficient use of available water supplies becomes more imperative. In areas where increased agricultural output is a prerequisite to more general economic expansion, water can be instrumental in bringing arid land into production and expanding current production levels. Water applied in wrong quantities, however, can generate salinity conditions if soil drainage is inadequate, can leach important plant nutrients beyond the plant root zone, or can depress yields if not used in proper proportions with other production inputs. The immediate consequence of any of these is an inefficient use of water, thus the need for more complete knowledge of plant-water-soil relationships.

Considerable interdisciplinary research into estimating production functions has been undertaken over the past two decades. While primary emphasis has been on conducting fertilizer and animal nutrition trials, research into plant-water-soil relationships has been gaining momentum in recent years. Several of these latter studies were reviewed in Chapter 3. Beringer has noted that the lesser number of water production studies is not due to water being unimportant but rather to a consequence of "the difficulty of translating certain concepts used by agronomists into terms that can be used immediately in economic analysis."[1] Some of these concepts were defined and discussed in Chapter 3. In addition, field experiments for studying plant-water-soil relationships are relatively costly.

Of central interest here are the analyses of experimental data generated by field experiments conducted in several western states. While these input-output data are restricted to a selected number of crops and only two inputs, water and fertilizer, knowledge of these plant-water-fertilizer-soil relationships has several applications. These relationships are important in developing optimal farm management programs. The rates of substitution among water, fertilizer, and land have direct application to efficient resource allocation. Optimal applications of water must, of course, be determined within the context of resource availabilities, competing uses, soil differences, and the specific production objectives being pursued. Evaluation of the economic feasibility of proposed irrigation projects requires data on input-output relationships for water. These relationships are particularly useful in predicting yields in previously nonirrigated areas. To the extent data are available, individual farm and project demand functions for water can be estimated.

As noted earlier, data on input-output relationships for the plant-water-fertilizer-soil complex are relatively scarce. Farm surveys generally provide point estimates. That is, producers can indicate an estimate of per acre yield associated with the quantity of water and fertilizer applied. Field experiments usually provide sufficient yield observations for estimating a continuous input-output relationship. With estimation of a continuous function, the following types of information are derivable: (1) predicted yields corresponding to specified input use-levels, (2) marginal physical products and marginal value products for each input, (3) marginal rates of substitution among inputs, (4) derived economic demand for individual inputs, (5) isoquants and isoclines for input-output relationships, (6) production elasticities and price elasticities of demand for inputs, and (7) product supply functions for varying price relationships and input use-levels.

Predicted yields corresponding to specified input use-levels provide technical input-output coefficients for analyses of optimal resource allocation for a farm or region of representative farms. As price relationships vary, estimates of the marginal value products of inputs and the marginal rates of substitution among inputs are integral to determining necessary changes in input-use patterns so that economic efficiency can be maintained. The other types of information generated are important for estimating demand-supply conditions in both the input and output markets.

The materials and analyses in this study were developed as an outgrowth of Contract 14-06-D-6192 between the Bureau of Reclamation of the U.S. Department of the Interior and the Center for Agricultural and Rural Development at Iowa State University. The generation of primary input-output data presented in this study represents a step toward expanding knowledge of water response relationships in agriculture so that more intelligent, more coordinated public programs of water use can be developed and implemented.

SELECTION OF CROPS AND EXPERIMENTAL SITES

Among the several crops considered for field experimentation, corn, wheat, sugar beets, and cotton were selected as the major crops having widest geographical applicability. Other crops such as vegetables, alfalfa, and rice are important regional crops. A study of regional economic development would need to include those crops that are locally important. The four crops included for experimentation were considered to be of currently or potentially major economic importance throughout much of western United States. Selection of experimental sites was conditioned by the nature of the cooperating agronomists' ongoing research programs. In most situations, the field experiments conducted for this study represented extensions and continuation of previous research programs. Knowledge generated from prior research was extremely useful in determining appropriate treatment levels for water and fertilizer and for general conduct of the experiments.

Individual field experiments generate input-output data relevant for the specific site and crop being considered. Variation across sites having different soil and environmental features is also of interest. The Yuma Valley and Yuma Mesa, Arizona, sites, for example, are about 10 miles apart. The soil features, however, are substantially different. Soil at the Yuma Mesa site is sandy while soil at the Yuma Valley location is of a clay loam texture. Consequently, the same crop grown at both sites has different water requirements, both in terms of irrigation scheduling and quantity applied. To capture this

TABLE 5.1. Specification of field experiments, by location and year

Year	Location	Year	Location
	Corn		Cotton
1968	Ft. Collins, Colorado	1967	Shafter, California
1969	Davis, California	1968	Shafter, California
1970	Davis, California	1969	Shafter, California
1969	High Plains, Texas	1967	West Side, California
1970	High Plains, Texas (2)[a]	1968	West Side, California
1971	High Plains, Texas (2)[a]	1969	West Side, California
1970	Colby, Kansas	1971	Yuma Valley, Arizona
1971	Colby, Kansas	1971	Yuma Mesa, Arizona
1970	Yuma Valley, Arizona	1971	Tempe, Arizona
1970	Yuma Mesa, Arizona	1971	Safford, Arizona
1970	Mesa, Arizona		
1971	Mesa, Arizona		
1970	Safford, Arizona		
1972	Safford, Arizona		
	Wheat		Sugar Beets
1969	Ft. Collins, Colorado	1969	Ft. Collins, Colorado
1970	Walsh, Colorado	1970	Walsh, Colorado
1970–71	Yuma Valley, Arizona	1969	High Plains, Texas
1971–72	Yuma Valley, Arizona	1970	High Plains, Texas
1970–71	Yuma Mesa, Arizona	1971	High Plains, Texas (2)[a]
1971–72	Yuma Mesa, Arizona	1969–70	Mesa, Arizona
1970–71	Mesa, Arizona	1970–71	Mesa, Arizona
1971–72	Mesa, Arizona	1971–72	Mesa, Arizona
1970–71	Safford, Arizona	1969–70	Yuma Valley, Arizona
	Corn Silage	1970–71	Yuma Valley, Arizona
		1969–70	Yuma Mesa, Arizona
1968	Ft. Collins, Colorado	1970–71	Yuma Mesa, Arizona
1970	Yuma Valley, Arizona	1970	Safford, Arizona
1970	Yuma Mesa, Arizona	1972	Safford, Arizona
1970	Mesa, Arizona		
1971	Mesa, Arizona		

[a]Experiments conducted at upland and lake sites.

TABLE 5.2. Individuals supervising, coordinating, and (or) conducting experiments at designated sites

ARIZONA:	Dr. Martin A. Massengale Head, Department of Agronomy and Plant Genetics The University of Arizona Tucson, Arizona
Mesa:	Mr. John M. Nelson (sugar beets) Mr. Rex K. Thompson (corn, wheat)
Safford:	Dr. Fred Turner, Jr.
Tempe:	Mr. Lloyd L. Patterson
Yuma Mesa *Yuma Valley*	Dr. Ernest B. Jackson
CALIFORNIA:	Dr. Robert M. Hagan Department of Water Science and Engineering University of California Davis, California
Davis:	Dr. J. Ian Stewart
Shafter *West Side*	Dr. Donald W. Grimes
COLORADO:	
Ft. Collins *Walsh*	Dr. Robert E. Danielson Department of Agronomy Colorado State University Ft. Collins, Colorado
KANSAS:	Mr. Evans E. Banbury Superintendent, Colby Branch Station Kansas Agricultural Experiment Station Colby, Kansas
Colby:	John R. Lawless
TEXAS:	Mr. Jim Valliant Director of Research High Plains Research Foundation Plainview, Texas

across-site variability, experiments were conducted at sites of differing soil and weather conditions. To make the study manageable and within the confines of financial constraints, one or more of the four major crops were studied at a total of thirteen experimental sites dispersed among five states. Cotton, for example, was not studied in Texas; wheat was not included in Kansas. The number and selection of crops at each experimental site were further conditioned by the degree to which these field experiments could be integrated with previous and existing research programs. The field experiments included in this study are specified in Table 5.1. Agronomists conducting and supervising the field experiments are listed in Table 5.2.

Experimental design

An "incomplete block design involving factorial treatments" was used for most field experiments. As noted in Chapter 4, the design is incomplete because all possible treatment combinations are not included. The factorial arrangement, that is, the specification and distribution of various treatment combinations, was designed to facilitate estimation of coefficients for a second-order polynomial production function. With this design, relatively more observations are available for fitting a function to the more important points on, for example, a quadratic surface, as depicted in Figure 4.3. This distribution of observations is also suitable for fitting square root, three-halves, and other similar polynomial forms to the experimental data. The design also provides sufficient points for a reasonably balanced goodness of fit test.

The basic factorial arrangement for each block is given in Figure 5.1. The treatment levels are coded so that, for example, $I(2) = 17.5$ inches of water and $F(3) = 175$ pounds of nitrogen fertilizer for the 1969 cotton experiment at Shafter, California. An X indicates that the corresponding treatment combination was replicated once per block; and 0 designates a single observation. Treatment combinations were randomly assigned to individual plots. Two blocks were used in most experiments. Plot size varied with the crop grown, soil characteristics, and method of irrigation. The appropriate plot size was determined by agronomists at each experimental site.

A "randomized complete block design with factorials" was used for the Texas experiments. The corn experiments at Davis, California, were set up to complement ongoing research into the impact of plant population on yield. Consequently, a "randomized incomplete block design with factorials and split plots" was used. Treatment combinations were randomly assigned to plots and then each plot was split with high and low plant population. Fi-

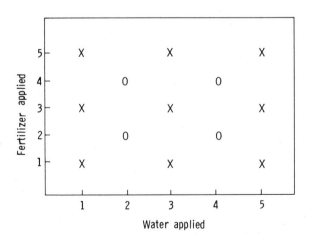

FIG. 5.1. Distribution of treatments within an incomplete block design with factorials, per block of experimental plots.

nally, the central composite rotatable design was used for the 1967 and 1968 cotton experiments in California.

The number of treatment replications and the number of experimental plots were determined by the size of the experimental area and financial constraints. Most experiments incorporated two blocks, each containing 22 experimental units (plots). In a few trials, three blocks were used. Blocking was not used in the Texas experiments.

Fertilizer treatments

Based on knowledge of plant-soil relationships and previous empirical experience, agronomists at each experimental site had the primary responsibility for determining appropriate treatment levels. Treatment levels were planned to cover a range sufficient to reproduce the yield relationships represented in Figure 4.3. Treatment levels were chosen to reflect residual soil nutrients from preceding experiments and current plant requirements. Measurements of preplant soil fertility were not made for each experiment. In most cases, sufficient quantities of phosphate and (or) potassium were applied prior to planting so that these plants and nutrients would not have a limiting effect on yield.

Irrigation treatments

Two basic procedures were used to determine the timing of irrigations and the quantity applied. For most experiments tensionmeters measuring soil moisture tension and, in turn, the need for an irrigation treatment were placed at appropriate positions in the soil. Soil moisture blocks and (or) neutron probe tubes were also used in determining the timing of irrigation treatments. Whenever soil moisture tension rose to a predetermined level, sufficient water was applied to restore soil moisture to the field capacity level. In the Texas experiments, irrigation scheduling was geared to the stages of plant growth. The quantity of water supplied was related to yield-water relationships generated in prior research.

In estimating the production functions in later chapters, the quantities of water applied at each irrigation have been aggregated to W, total water applied during the growing season. Consequently, a specific irrigation scheduling regime is implicit in each model containing W. The irrigation scheduling programs are not reproduced here but are obtainable elsewhere.[2]

Following planting and any irrigations for germination, experimental plots were bordered and basined to minimize water losses through runoff. Appropriate quantities of water were metered into the basins during the irrigations. Rainfall was a factor in several experiments. As a rule of thumb, only that rainfall exceeding 0.25 inch at any occurrence during the growing season was included in the analyses. For some experiments, available soil moisture at harvest was measured. Deducting this quantity from the beginning level of available soil moisture plus water applied plus rainfall, estimates of total water available for plant growth plus water losses through evaporation and percolation were derivable.

DATA GENERATED FOR ANALYSIS

Data on yield-water-fertilizer relationships for individual corn, wheat, cotton, and sugar beet experiments are given in subsequent chapters. Summaries of the data on soils characteristics at the respective sites are given in Appendix Tables 1 and 2. To limit the length of this study, climatological data for the individual experiments are not detailed. Levels of rainfall are reported.

ANALYSIS OF CORN EXPERIMENTS

Sixteen field experiments for corn (grain) were conducted at nine experimental sites. The number of repetitions at each site was partly influenced by the cooperating agronomist's ongoing research programs. Locations of the experiments were listed in Table 5.1. Since field trials are always subject to exogenous forces beyond the control of the agronomist, particularly weather, the quality of the experiments is somewhat variable.

The experimental data are presented for each experiment discussed in this chapter. To the extent available, supplementary information on growing conditions is provided. Planting and harvesting dates and data on soil characteristics are summarized in the appendix tables. Several variations of independent variables are used in subsequent analyses. Let W represent acre-inches of water applied after planting. W_1 is defined as water applied, W, plus total rainfall exceeding 0.25 inch at any occurrence during the growing season plus either water applied during a preplant irrigation or an estimate of available soil moisture (ASM) at planting. Estimates of ASM at harvest are available for a few experiments. When ASM at harvest is subtracted from W_1, an estimate of water used by the plant plus water losses through soil evaporation and percolation is derived. This water variable is designated W_2. For fertilizer, N denotes pounds of nitrogen applied per acre while N_1 embodies N plus estimates of preplant soil nitrogen available for plant growth. Data for W_2 and N_1 are not available for all experiments. Since W_1 and N_1 are most closely related to plant growth and yields, analyses will focus on these two factors and to a lesser extent on W, W_2, and (or) N.

Each set of experimental data is routinely subjected to an analysis of variance to determine occurrence of a block effect or plant population effect. If no block effect is observed, the soil characteristics and management practices among blocks are assumed essentially homogenous, and the experimental data from individual blocks are combined. Plant population is similarly treated. When a statistically significant block effect is observed at the 5 percent level, dummy variables for each block are incorporated into the model.[1]

The commonly used quadratic, square root, and three-halves polynomials were routinely "fitted" to the experimental data. These forms have properties suitable for estimating plant-water-fertilizer relationships. Since these functions do not vary substantially in terms of test statistics, only one of the three will be specified for discussion. In selected instances, an exponential function will also be fitted to the data. Levels of statistical significance for the derived coefficients are designated as *** (1 percent), ** (5 percent), and * (10 percent). These are occasionally referred to as the "conventional" levels of statistical significance. The standard errors of the derived coefficients are given in parentheses.

The "lack of fit" F test was also conducted for each production function. If the test indicated that the functional form did not adequately fit the data being analyzed, the $(Y - \hat{Y})$ residuals were examined for correspondence with the assumptions implicit in ordinary least-squares procedures for deriving production functions. This procedure was described in Chapter 4. Routine deletion of observations, however, is not a good practice; the analyst can "force" the results. When observations are omitted to improve the functional form, the data deleted and the resulting analysis are fully specified.

Following selection of the most appropriate production function among those routinely considered, a table of predicted yields is derived by substituting alternative combinations of water and fertilizer use-levels into the function. Information such as maximum yield, yield response functions to water and fertilizer, and average physical products for water and fertilizer is derivable from this table. The 1971 experiment at Colby, Kansas, is analyzed in substantial detail to demonstrate the types of information obtainable from field experiments. Time and space prohibit this more detailed analysis of each experiment.

COLBY, KANSAS, 1971

Prairie Valley 40–S hybrid was planted on Keith Silt Loam at the Colby Branch Experiment Station. The level of soil moisture at planting was at or near field capacity. The estimated available water-holding capacity (AWC) of this soil is 8.1 acre-inches in the top 4 feet. Neutron probes were installed in each of the 44 experimental plots to measure soil moisture. Five irrigation regimes required that designated plots be irrigated when ASM was 20, 35, 50, 65, and 80 percent with sufficient water applied at each irrigation to return soil moisture to the field capacity level. Since plots were individually measured for soil moisture levels, those subjected to the same water treatment received differing amounts of water. For example, water applications on the 4 plots receiving no nitrogen treatment and irrigated when ASM fell to 20 percent ranged from 8.32 to 9.60 acre-inches.

Soil samples from individual plots were analyzed for levels of preplant nitrogen. Plot measurements ranged from 20.2 to 72.7 pounds, averaging 42.0 pounds per acre. The five nitrogen treatments ranged from 0 to 360 pounds per acre, in increments of 90 pounds.

Two blocks, each with 22 plots, were incorporated into a randomized incomplete block experimental design. Because water treatments varied with each plot, treatment combinations were not replicated. Environmental conditions during the growing season were normal. Rainfall exceeding 0.25 inch at any occurrence totaled 7.46 inches from plant to harvest. Treatment levels and corresponding yields are listed in Table 6.1.

No block or plant population effect is discernible at conventional levels of statistical significance. The quadratic form in Equation (6.1) represents yield-water-fertilizer relationships for this experiment.

$$\text{yield} = -1354.4861 + 434.0799^{***}W + 39.9816^{***}N - 11.0341^{**}W^2$$
$$(1078.40) \quad\quad (145.07) \quad\quad (3.496) \quad\quad (4.507)$$
$$- .0840^{***}N^2 + .3874^{**}WN \quad\quad\quad\quad\quad\quad\quad (6.1)$$
$$(.008) \quad\quad (.169)$$
$$R^2 = .936 \quad\quad F = 111.82^{***}$$

Since the water treatment levels were not replicated, the *LOFF* test cannot be made. When the $(Y - \hat{Y})$ residuals are plotted against W and N, all residuals associated with $N = 90$ are considerably above the zero line. When these four observations are deleted, the quadratic form is reestimated as (6.2). Compared to (6.1), the level of significance for several coefficients has decreased.

$$\text{yield} = -3040.3061 + 684.0004^{***}W + 36.2528^{***}N - 19.9511^{***}W^2$$
$$(1077.35) \quad\quad (152.30) \quad\quad (3.255) \quad\quad (5.095)$$
$$- .0803^{***}N^2 + .6023^{***}WN \quad\quad\quad\quad\quad\quad\quad (6.2)$$
$$(.008) \quad\quad (.182)$$
$$R^2 = .957 \quad\quad F = 149.65^{***}$$

Plant growth is not only affected by the quantities of water and fertilizer applied but by preplant soil moisture and fertility and by rainfall during the growing season. When W_1 is defined as water applied plus 8.1 acre-inches preplant available soil moisture plus 7.46 inches rainfall and N_1 as nitrogen applied plus estimated preplant soil nitrogen for respective plots, the quadratic form in (6.3) is generated.

$$\text{yield} = -10586.0287 + 688.3554^{**}W_1 + 36.4211^{***}N_1 - 10.0386^{**}W_1^2$$
$$(4559.14) \quad\quad (296.83) \quad\quad (5.344) \quad\quad (4.756)$$
$$- .0772^{***}N_1^2 + .4133^{**}W_1N_1 \quad\quad\quad\quad\quad\quad\quad (6.3)$$
$$(.008) \quad\quad (.170)$$
$$R^2 = .930 \quad\quad F = 100.32^{***}$$

Since levels of W_1 and N_1 vary for each experimental unit, plots of $(Y - \hat{Y})$ against W_1 and N_1 are not meaningful. The statistical properties of (6.3) are reasonably similar to those estimated for (6.1). The variables in (6.3), however, are more closely related to determinants of plant growth and yield.

Estimates of W_2 are reported in Table 6.1a. The quadratic form incorporating W_2 is specified in (6.4).

$$\text{yield} = -13485.5330 + 1033.2758^{***}W_2 + 23.1481^{***}N_1 - 17.5419^{**}W_2^2$$
$$(4841.10) \quad\quad (379.96) \quad\quad (8.104) \quad\quad (7.301)$$
$$- .0735^{***}N_1^2 + .7743^{**}W_2N_1 \quad\quad\quad\quad\quad\quad\quad (6.4)$$
$$(.009) \quad\quad (.311)$$
$$R^2 = .928 \quad\quad F = 98.44^{***}$$

Compared to (6.3), coefficients for the intercept and W_2 have a slightly lower level of statistical significance. The question arises as to whether or not (6.4) is more appropriate than (6.3). To some extent the models are measuring different phenomena. W_2 is appropriate in an *ex post* context. That is, W_2 is a more exact measurement of the actual yield-water relationship. W_1, however, is of importance to the producer in that W_1 must be applied to generate W_2. Since $W_1 \geqq W_2$, the producer would prefer to apply W_2 less 7.46 inches rainfall, which would imply that ASM at harvest equals zero. If, however, irrigation treatments were designed to approximate this condition, yields would be reduced.

A five-parameter Mitscherlich model of the form in (3.8) and (3.9) was also fitted to these experimental data. Variable W' is defined as $(w = 8.1) + (W =$ water applied plus 7.46 inches rainfall). Recall that w represents "available soil moisture in acre-inches prior to first irrigation treatment." The quantified Mitscherlich formulation is in (6.5) where all coefficients except for W' are statistically significant at the 1 percent level. The maximum yield is estimated as 16,028 pounds per acre. In comparison, the maximum yield derived from (6.3) is much lower at 9985 pounds per acre. A negative value was estimated for c_1 and a positive value for k_1. A negative value for c_1, that is, a negative "effect factor" for water is not meaningful; the coefficient for k_1 should be negative. Recall that the k_1 and k_2 coefficients associated with the quadratic variables permit yield to reach a maximum and then decline. The estimated R^2 for (6.5) was 0.945.

$$\text{yield} = 16\overset{***}{0}27.9 \; \left[1 - e^{(.0590)} \; \overset{.0937W' \; -.00\overset{***}{4}7W'^2}{e^{(.0021)}}\right]$$
$$(3562.72)$$
$$\left[1 - e^{(.0018)} \; \overset{-.00\overset{***}{5}6N_2 \; .0000\overset{***}{0}9N_2^2}{e^{(.000003)}}\right] \tag{6.5}$$

Equation (6.3) is used to generate several other physical and economic relationships of interest to producers and economists. Predicted yields derived from (6.3) are listed in Table 6.2. Selecting any column, yield = $f(W|N)$ represents a water response function for the specified level of N. By selecting any row in Table 6.2, results of a nitrogen response function are observable where yield = $g(N|W)$. The yield response to nitrogen is especially strong when N is increased in the range of 60 to 120 pounds per acre. When 44 acre-inches of water are available, yields are decreasing at all levels of nitrogen. The maximum yield associated with (6.3) is derived as 9985 pounds per acre generated when about 41.4 acre-inches of water and 347 pounds of nitrogen per acre are available. Recall the maximum physical product is determined by setting the marginal products of water and fertilizer derived from (6.3) equal to zero, solving the two equations simultaneously for W_1 and N_1, and then substituting these values into (6.3).

A three-dimensional production surface based on points estimated from (6.3) is plotted in Figure 6.1. This production surface includes all the yields listed in Table 6.2. Curve AB, for example, represents the yield response to

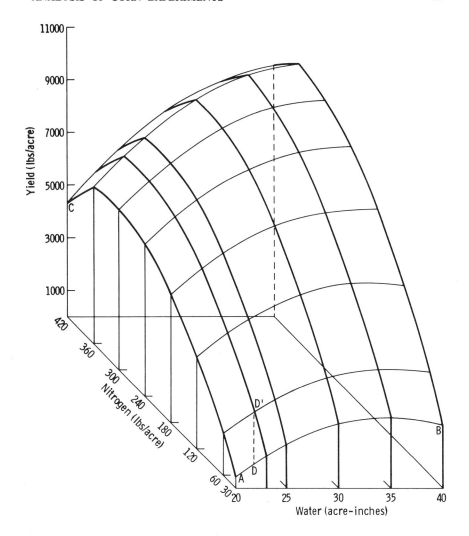

FIG. 6.1. Production surface for corn (grain) estimated
from (6.3) according to specified water and
nitrogen Levels (Colby, Kansas, 1971).

water when nitrogen availability is 30 pounds per acre. Curve AC reflects the
yield response to successively larger applications of nitrogen when 20 acre-
inches of water are available. The other curves are similarly interpreted.
When $W = 23$ and $N = 60$, the distance DD' to the frontier of the production
surface represents 1976.6 pounds per acre of corn grain, the predicted yield
for these levels of water and fertilizer. If the available soil nitrogen prior to
planting was 30 pounds per acre, an additional 30 pounds would need to be
applied to produce a yield of 1977 pounds per acre.

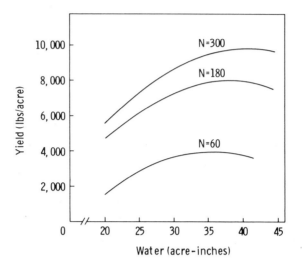

FIG. 6.2. Yield response curves for corn (grain) to water derived from (6.3) assuming specified levels of nitrogen (Colby, Kansas, 1971).

Yield response curves to water when N is fixed at 60, 180, and 300 pounds per acre are in Figure 6.2. For $N = 60$, the response curve reaches a maximum when $W = 35.5$ acre-inches. The marginal product of water is zero at this point and becomes negative with larger applications of water. When N is increased to 180 and 300 pounds per acre, the marginal product of water, MP_W, is zero when $W = 38.0$ and 40.46 acre-inches, respectively. For any level of water, the vertical distance between the curves represents the yield generated when an additional 120 pounds of N per acre are applied. Note that at higher levels of fertilization, the distance between the curves diminishes. Yield response curves to nitrogen are plotted in Figure 6.3. When $W = 44$ and the level of fertilization is relatively low, too much water is available and yields are depressed.

Marginal product curves derived from the quadratic form are linear. The equation for MP_W, for example, is given in (6.6) where the level of N determines the relative position of the MP_W curve.

$$MP_W = \partial(\text{yield})/\partial W_1 = 688.3354 - 20.0772W_1 + .4133N \qquad (6.6)$$

At each level of N, however, MP_W declines at a constant rate of -20.0772 for each incremental acre-inch of water. In Figure 6.5, MP_N declines at a constant rate of -0.1544 for each level of water availability. Marginal product curves for water and nitrogen are plotted in Figures 6.4 and 6.5.

So far only water and fertilizer have been introduced as variable inputs. Land is also a possibility. If specific quantities of water and fertilizer are assumed available, these can be applied rather intensively on a few acres of land

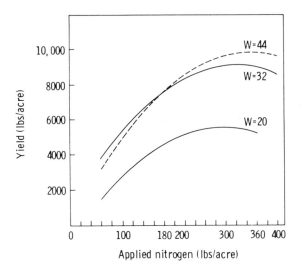

FIG. 6.3. Yield response curves for corn (grain) to nitrogen derived from (6.3) assuming specified levels of water applications (Colby, Kansas, 1971).

or spread over a relatively large area. To examine these possibilities, a land variable designated as A is incorporated in (6.3). The other variables are redefined as yield* = yield $\times A$, $W^* = W \times A$ and $N^* = N \times A$ and (6.3) is rewritten as (6.7):

$$\text{yield}^*A^{-1} = -10586.0287 + 688.3354W^*A^{-1} + 36.4211N^*A^{-1}$$
$$- 10.0386(W^*A^{-1})^2 - .0772(N^*A^{-1})^2 \qquad (6.7)$$
$$+ .4133W^*N^*A^{-2}$$

$$\text{yield}^* = -10586.0287A + 688.3554W^* + 36.4211N^*$$
$$- 10.0386(W^*)^2A^{-1} - .0772(N^*)^2A^{-1} \qquad (6.8)$$
$$+ .4133W^*N^*A^{-1}$$

Subscripts in (6.7) and (6.8) have been omitted. Equations (6.3) and (6.7) are identical statements. Land is introduced as a variable by multiplying (6.7) by A, thereby generating (6.8). Yield* is now a function of W^*, N^*, and acres of land. The marginal product of the land function, MP_A, is given in (6.9) and is formulated as the first derivative of yield* in (6.8) with respect to A. Three levels of W^* and N^* for application on successively larger areas of land have been assumed. The three levels of water and nitrogen availability are 750, 1000, and 1500 acre-inches and 2500, 5000, and 7500 pounds, respectively. The MP_A curves in Figure 6.6 are curvilinear because of the A^2 term in (6.9).

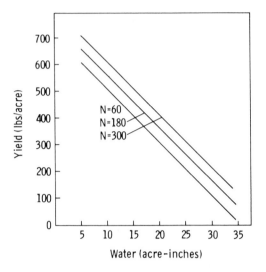

FIG. 6.4. Marginal product curves for water in production
of corn (grain) at specified levels of nitrogen
(Colby, Kansas, 1971).

$$\partial(\text{yield}^*)/\partial A = -10586.0287$$

$$+ [10.0386\,(W^*)^2 + .0772\,(N^*)^2 - .4133W^*N^*]/A^2 \qquad \textbf{(6.9)}$$

The concept and importance of isoquants and isoclines were discussed in
Chapter 2. The isoquant equation, derived from (6.3), is specified in

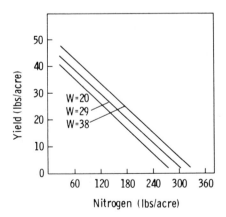

FIG. 6.5. Marginal product curves for nitrogen in the
production of corn (grain) at specified levels of
water (Colby, Kansas, 1971).

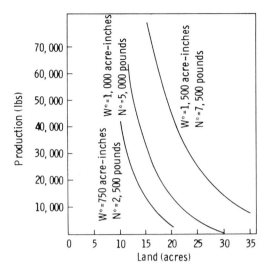

FIG. 6.6. Marginal product curves for land in production of corn (grain) with specified availability of water and nitrogen (Colby, Kansas, 1971).

(6.10). The equation is written with water, W, as the dependent variable; it also could have been rewritten in terms of nitrogen, N. Subscripts have been omitted. After selecting a yield level, that is, the value of the isoquant, and a level for N, the corresponding level of W necessary to produce that yield is then derived. This combination of W and N represents one point on the isoquant being plotted. A family of isoquants is estimated by assuming different levels for yield and deriving combinations of W and N that will generate these respective yields. Isoquants for three yield levels and isoclines for three price relationships are plotted in Figure 6.7.

$$W = \{ 688.3354 + .4133N \pm [(-688.3354 - .4133N)^2$$

$$- (4)(10.0386)\ (\text{yield} + 10586.0287 - 36.4211N \qquad \textbf{(6.10)}$$

$$+ .0772N^2)]^{.5}\}/2(10.0386)$$

The marginal rate of substitution between W and N at any point on an isoquant is determined by the first derivative of (6.10) with respect to N and then substituting the values of yield, W, and N corresponding to the point being considered. Consider point A in Figure 6.7. The slope of isoquant $Y = 4000$ at point A is 6. This follows since the marginal rate of substitution of nitrogen for water, MRS, is equal to the price of water, P_w, divided by the price of nitrogen, P_n, or $MRS = P_w P_n^{-1} = 6$ at the point where isocline I_2 intersects isoquant $Y = 4000$. When P_w increases relative to P_n, for example, I_3, substitution of fertilizer for water becomes profitable and the point representing the profit-maximizing combination of W and N shifts from A to B.

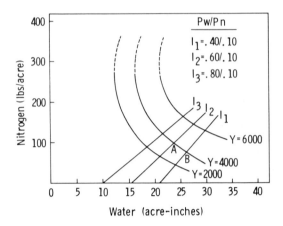

FIG. 6.7. Isoquants and isoclines for specified per acre
yield of corn (grain) and indicated price ratios
(Colby, Kansas, 1971).

In addition to deriving the *MRS* of *N* for *W*, substitution possibilities also exist between *W*, *N*, and land. The *MRS* between land and *W*, $\partial A/\partial W$, is the ratio of MP_W to MP_A or $[\partial(\text{yield*})/\partial W^*]/[\partial(\text{yield*})/\partial A]$, as in (6.10') where *B* is represented by (6.3). The *MRS*'s between land and *N* and between *W* and *N* are similarly derived. A limited number of *MRS*'s are listed in Table 6.3.

$$\partial A/\partial W = (688.3354A - 20.0772AW + .4133AN)/B \qquad \textbf{(6.10')}$$

COLBY, KANSAS, 1970

The experimental design and treatment levels were essentially identical to the 1971 experiment. As with the 1971 experiment, water treatments were not replicated but varied with individual plots. Rainfall exceeding 0.25 inch at any occurrence totaled 11.87 inches during the growing season.

Measurements of preplant soil fertility were not made for this experiment. In the two previous years, experiments at this site were treated to applications of 200 pounds of nitrogen per acre. Assuming a carry-over rate of 20 percent, residual *N* is estimated at 40 pounds per acre. Preplant *N* in the 1971 experiment averaged 42 pounds per acre across the 44 plots. Sufficient phosphorus was applied to bring plot levels up to 54 pounds of total phosphorus per acre. Treatment combinations were randomly assigned to two blocks, each containing 22 plots. Treatment levels and corresponding yields are given in Table 6.4. Based on an analysis of variance, there was no block effect but a plant population effect was observed for this experiment. Estimates of plant population were made at harvest.

Yield-water-nitrogen relationships for this test were represented by the 1.5 polynomial in (6.11). Plant population, expressed in terms of 10,000 plants per acre, is denoted by P. Since the water treatments were not replicated, a *LOFF* test cannot be made. The $(Y - \hat{Y})$ residuals, however, were plotted against W_1 and N_1. The residuals for the four observations associated with $N = 270$ are all substantially below the zero line.

$$
\text{yield} = -25615.2995^{**} + 3443.8448^{**}P + 1887.0266^{*}W_1
$$
$$
(11636.89) \qquad (1342.76) \qquad (977.64)
$$
$$
+ 18.5261^{**}N_1 - 201.6017^{*}W_1^{1.5} - 1.0360^{***}N_1^{1.5}
$$
$$
(8.089) \qquad (108.76) \qquad (.276)
$$
$$
+ .2565W_1N_1
$$
$$
(.177)
$$
$$
R^2 = .791 \qquad F = 23.28^{***}
$$

(6.11)

After omitting these observations, the 1.5 polynomial is reestimated as

$$
\text{yield} = -31265.2852^{**} + 3375.8053^{**}P + 2381.4411^{**}W_1
$$
$$
(12198.29) \qquad (1447.50) \qquad (1024.55)
$$
$$
+ 18.0662^{**}N_1 - 257.4470^{**}W_1^{1.5} - 1.1370^{***}N_1^{1.5}
$$
$$
(8.157) \qquad (114.08) \qquad (.283)
$$
$$
+ .3406^{*}W_1N_1
$$
$$
(.184)
$$
$$
R^2 = .799 \qquad F = 21.91^{***}
$$

(6.12)

All estimated coefficients are statistically significant at the 10 percent or lower level. Within the range of observations, a positive relationship exists between yield and plant population. Interaction terms such as PW_1 and PW_1N_1 were also considered in other models. The results, however, were not superior to those in (6.12). Predicted yields derived from (6.12) with an assumed average plant population of 17,300 plants per acre are listed in Table 6.5.

Yield-water-nitrogen relationships can also be quantified in terms of water used, W_2. Estimates of W_2 are in Table 6.4. When W_2 is incorporated into an analysis of variance, both block and plant population effects are statistically significant at the 5 percent level. The estimated 1.5 polynomial is expressed in terms of W_2 as in (6.13) where X_1 represents the block effect. All coefficients except for N_1 are statistically significant at the 10 percent or lower level. Based on plots of residuals against W_2 and N_1, all residuals generated by the $N = 270$ observations fell below the zero line. When these observations are deleted, however, the reestimated 1.5 polynomial is not superior to (6.13).

$$\text{yield} = -24364.6010^{**} + 440.3723^{**}X_1 + 3181.3043^{**}P$$
$$(11018.91) \qquad (208.18) \qquad (1294.42)$$

$$+ 2225.8490^{*}W_2 + 7.3305N_1 - 263.3108^{*}W_2^{1.5}$$
$$(1168.78) \qquad (9.140) \qquad (145.30) \qquad\qquad (6.13)$$

$$- .9557^{***}N_1^{1.5} + .6135^{**}W_2N_1$$
$$(.273) \qquad\quad (.262)$$

$$R^2 = .812 \qquad F = 22.26^{***}$$

If a predicting equation is required for Block 2, $X_1 = 0$ and the intercept for the prediction model is -24364.60. For Block 1, $X_1 = 1$ and the intercept becomes $-24364.60 + 440.37 = -23924.23$. The other variables and coefficients are unchanged.

FT. COLLINS, COLORADO, 1968

Kitley K4–117 was planted in 28-inch rows on Nunn Clay Loam at the Agronomy Research Center. Plant population approximated 25,000 per acre. The observed plot yields for corn grain and silage corresponding to specified treatment levels are summarized in Table 6.6. Irrigation treatments $I_1, \ldots,$ I_5 reflect the maximum allowable soil moisture tension levels of 9, 6, 3, 1, and 0.7 bars, respectively. Compared to a 70-year average, monthly temperatures during the growing season averaged close to normal while total precipitation was about 1.5 inches below the long-term average. Rainfall during the growing season totaled 5.13 inches.

Corn grain

Based on an analysis of variance, a "block effect" was implicit in the data. Consequently, dummy variables X_1 and X_2 were included for block identification where

$$X_1 \quad X_2$$

$$1 \quad 0 \quad = \text{Block 1}$$

$$0 \quad 1 \quad = \text{Block 2}$$

$$0 \quad 0 \quad = \text{Block 3}$$

The block effect may have been due to soil heterogeneity but more likely to differences in preplant or residual nitrogen (NO_3) among the three blocks. Levels of NO_3 in the top 3 feet of soil were estimated at 106, 45, and 269 pounds per acre for Blocks 1, 2, and 3, respectively.

In fitting alternative polynomial forms to the 66 yield observations identified by block, the 1.5 polynomial seemed statistically appropriate. The derived production function is given in (6.14) where the independent variables are as earlier defined. Three predicting equations are derivable from (6.14), one for each block. The model for any one block is abstracted by substituting the values of the dummy variables corresponding to that block.

$$\text{yield} = -14327.1899^{***} + 887.1194^{***}X_1 + 1671.1632^{***}X_2$$
$$(5050.38) \qquad (342.67) \qquad (385.42)$$

$$+ 2522.1524^{***}W_1 + 15.6510^{**}N_1 - 354.2371^{***}W_1^{1.5}$$
$$(820.17) \qquad (7.677) \qquad (127.18)$$

$$- .6519^{**}N_1^{1.5} + .2139W_1N_1 \qquad\qquad (6.14)$$
$$(.268) \qquad (.241)$$

$$R^2 = .700 \qquad F = 19.35^{***} \qquad LOFF: \text{Block } 1 = 1.599$$
$$\text{Block } 2 = .785$$
$$\text{Block } 3 = .236$$

For comparison, the 1.5 polynomial excluding the block effect is in (6.15). The estimated coefficients are almost of the same magnitude as those in (6.14) but with higher standard errors.

$$\text{yield} = -12484.6811^{**} + 2506.7492^{**}W_1 + 13.3899^{***}N_1 - 352.0218^{**}W_1^{1.5}$$
$$(5725.28) \qquad (933.25) \qquad (8.488) \qquad (144.71)$$

$$- .7407^{**}N_1^{1.5} + .2189W_1N_1 \qquad\qquad (6.15)$$
$$(.286) \qquad (.274)$$

$$R^2 = .598 \qquad F = 17.87^{***} \qquad LOFF = 1.345$$

Using the model specified in (6.14) but redefining W_1 as W_2, the modified production function is in (6.16).

$$\text{yield} = -3527.9235 + 877.0275^{**}X_1 + 1666.7544^{***}X_2$$
$$(9234.35) \qquad (342.17) \qquad (384.80)$$

$$+ 849.9737W_2 + 15.5801^{*}N_1 - 84.0671W_2^{1.5}$$
$$(2159.20) \qquad (8.043) \qquad (398.79) \qquad (6.16)$$

$$- .6786^{**}N_1^{1.5} + .3416W_2N_1$$
$$(.268) \qquad (.373)$$

$$R^2 = .701 \qquad F = 19.44^{***}$$

Coefficients for the water variables are not statistically significant at conventional levels. The estimated standard errors are high relative to values of the derived coefficients. The R^2 is similar to that for (6.14); the $LOFF$ statistic was not calculated for individual blocks. The relationship between total moisture available, W_1, and total moisture used, W_2, for specified treatment levels is

	W_1 (acre-inches)	W_2 (acre-inches)	W_2/W_1
I_1	22.96	15.83	0.689
I_2	20.01	14.76	0.738
I_3	19.28	14.87	0.771
I_4	16.89	12.57	0.744
I_5	14.13	10.48	0.742

Variations of the Mitscherlich model were also estimated. The computational complexity of estimating this exponential form precludes including a block effect. Using the three-parameter model earlier specified in (3.9), the quantified version is in (6.17) where $N' = N + (n =$ an average of 140 pounds of preplant soil nitrogen). The maximum yield is estimated as 11,123.5 pounds or 198.6 bushels per acre. The growth factors for water and fertilizer are estimated to be 0.0795 and 0.0136, respectively. All derived coefficients were highly significant.

$$\text{yield} = 1112\overset{***}{3}.5 \left[1 - e^{\overset{-.079\overset{***}{5}W_1}{(.0152)}}\right] \left[1 - e^{\overset{-.01\overset{***}{3}6N'}{(.0019)}}\right]$$

$$R^2 = .619 \qquad LOFF = 1.352 \tag{6.17}$$

We use (6.14) as the most appropriate estimate for the yield data generated by this experiment. Predicted yields derived from (6.14) are listed in Table 6.7. Somewhere in the vicinity of $W = 24$, yields begin decreasing at all levels of fertilization.

Corn silage

The discussion of experimental conditions and procedures for corn grain also applies to corn silage. Corn for grain and silage was harvested from the same experimental units. A block effect was also observed in analyses of silage yields. The quadratic function in (6.18) provided a good fit for these data. Few of the estimated coefficients for the treatment variables, however, are statistically significant at conventional levels. Predicted yields derived from (6.18) are in Table 6.8.

$$\text{yield} = 14627.671\overset{*}{7} + 3475.3953X_1 + 7381.95\overset{***}{5}8X_2$$
$$(8329.61) \qquad (2356.91) \qquad (2664.83)$$

$$+ 2606.096\overset{*}{6}W_1 + 9.0504N_1 - 84.1496W_1^2$$
$$(1525.84) \qquad (28.76) \qquad (77.53)$$

$$- .0301N_1^2 + 2.936\overset{*}{2}W_1N_1 \tag{6.18}$$
$$(.046) \qquad (1.674)$$

$$R^2 = .588 \qquad F = 11.85^{***} \qquad LOFF: \text{Block 1} = 1.126$$
$$\text{Block 2} = 1.026$$
$$\text{Block 3} = 1.836$$

DAVIS, CALIFORNIA, 1970

Following a preplant irrigation of 4 inches, Pioneer 3775 was planted in 30-inch rows on Yolo Loam at the University of California Agricultural Experiment Field Station at Davis. A split plot design with factorial treatments was used. For each treatment combination randomly assigned to the experimental plots, the plot was split, with one half characterized by "high" plant population and the other half a "low" population. The planned "high" and "low" populations were 25,000 and 17,500 plants per acre. Based on previous research, a plant population near 25,000 per acre was considered optimal for high yields. The "low" plant population was introduced to examine the level of yield relative to the "high" population subplots when water and fertilizer were at low levels. The plant stands were thinned in mid-June. Actual plant populations for the "high" treatment ranged from 23,000 to 26,000 plants per acre among the various subplots. Plant population for the "low" treatment ranged from 15,500 to 21,000 plants per acre.

Treatment levels, plant population, and corresponding yields are summarized in Table 6.9. Referring to the average yields in Table 6.9, when $N = 0$, average yields from plots with "low" plant population were higher than those from the "high" population plots. When $N = 150$ and $W = 1.66$, the "low" population plots outyielded the "high" population ones. At higher levels of water and (or) nitrogen, a higher plant population can be sustained, and yields corresponding to the "high" population treatment were above those generated by the "low" population treatment.

Because of the split-plot design, a plant population effect was hypothesized for this set of data. The square root function in (6.19) was fitted to the data where P = plant population (10,000 per acre). The coefficient estimated for P is not statistically significant at conventional levels. When the plant population variable is omitted, the square root function is reestimated as (6.20). Using (6.20) as a predicting equation, estimated yields at various levels of water and nitrogen are in Table 6.10. The yield response to increased nitrogen applications is not strong at all water levels. Yields begin decreasing when the rate of fertilization is at a point between 90 and 150 pounds per acre. The yield response to water is considerably stronger, especially at relatively low levels of application.

$$\text{yield} = -8472.4401^{***} + 279.2162P - 577.5703^{***}W_1 - 7.2690^{**}N$$
$$\qquad\quad (1700.65) \qquad (390.32) \qquad (107.54) \qquad (3.612)$$

$$+ 6181.8664^{***}W_1^{.5} + 106.8208N^{.5} + 10.1050W_1^{.5}N^{.5} \qquad \textbf{(6.19)}$$
$$\quad (875.57) \qquad\qquad (79.30) \qquad\qquad (12.57)$$

$$R^2 = .930 \qquad F = 70.88^{***} \qquad LOFF = 1.482$$

$$\text{yield} = -8772.3648^{***} - 620.4960^{***}W_1 - 7.9160^{**}N + 6646.6751^{***}W_1^{.5}$$
$$(1675.70) \qquad (109.03) \qquad (3.731) \qquad (884.39)$$

$$+ 132.0856N^{.5} + 1.4460W_1^{.5}N^{.5} \qquad\qquad\qquad (6.20)$$
$$(81.91) \qquad\quad (13.08)$$

$$R^2 = .923 \qquad F = 78.82^{***} \qquad LOFF = 1.863$$

DAVIS, CALIFORNIA, 1969

The 1969 experiment is essentially identical to the one conducted in 1970. The growing season was described as "normal" with no significant amounts of rainfall recorded. Treatment levels, plant population, and corresponding yields are in Table 6.11. Average yields are also listed according to treatment combination and level of plant population.

Compared to the 1970 experiment, the "high" and "low" population treatments have a generally stronger effect. That is, the difference between average yields generated by the two plant populations for most treatment combinations tends to be greater than the corresponding differences in the 1970 experiment. Based on an analysis of variance, a plant population effect was observed. Models were fitted to data from the "low" population grouping, the "high" population, and the aggregated set of data with a plant population variable added. For the "low" plant population group, the quadratic form in (6.21) was estimated. Equation (6.22) represents yield-water-nitrogen relationships for observations from the "high" population group. In both (6.21) and (6.22), coefficients for the nitrogen variables are not statistically significant at conventional levels.

$$\text{yield} = 3297.4220^{***} + 367.1646^{***}W_1 + .5194N - 7.0578^{***}W_1^2$$
$$(426.04) \qquad (51.86) \qquad (3.397) \qquad (1.514)$$

$$+ .0038N^2 - .0458W_1N \qquad\qquad\qquad (6.21)$$
$$(.010) \qquad (.091)$$

$$R^2 = .910 \qquad F = 36.27^{***} \qquad LOFF = 0.564$$

$$\text{yield} = 1689.8893^{***} + 565.3938^{***}W_1 + 6.0766N - 10.6658^{***}W_1^2$$
$$(485.88) \qquad (59.14) \qquad (3.874) \qquad (1.727)$$

$$+ .0027N^2 - .1880^{*}W_1N \qquad\qquad\qquad (6.22)$$
$$(.011) \qquad (.103)$$

$$R^2 = .947 \qquad F = 63.71^{***} \qquad LOFF = .592$$

Referring to Table 6.11, the yield response to nitrogen at specific levels of water and plant population is sometimes increasing, decreasing, or both. Consequently, a consistent impact of nitrogen on yield is not discernible.

When data from the "low" and "high" populations are pooled, the square root function in (6.23) is generated. The estimated coefficient for P (population in terms of 10,000 plants per acre) is highly significant as are those for the intercept and water variables. Predicted yields derived from (6.23), assuming an average plant population of 22,000 per acre, are listed in Table 6.12. Since the coefficients for both nitrogen variables in (6.23) are positive, the marginal product of nitrogen increases indefinitely when water is fixed at any level and the rate of fertilization is successively increased. This phenomenon is also reflected in Table 6.12. Predicted yields are continually increasing when larger amounts of nitrogen are applied at any specific level of water.

$$\text{yield} = -4230.4644^{***} + 1251.8159^{***}P - 293.4513^{***}W_1 + 3.6167N$$
$$(1130.33) \qquad (262.04) \qquad (77.58) \qquad (3.075)$$

$$+ 3450.1654^{***}W_i^{.5} + 23.5011N^{.5} - 10.9958W_i^{.5}N^{.5} \qquad (6.23)$$
$$(574.24) \qquad (65.39) \qquad (10.30)$$

$$R^2 = .896 \qquad F = 58.74^{***} \qquad LOFF = 1.317$$

MESA, ARIZONA, 1971

Experimental plots at the Mesa, Arizona, site were treated to preplant applications of 250 pounds per acre of 11-48-0 fertilizer and 8 acre-inches of water, the latter raising the soil moisture to or near field capacity. Funks G-4949 was planted in 40-inch rows on Laveen Clay Loam. Plant population was approximately 20,000 per acre.

Irrigation scheduling and quantities applied were based on percent soil moisture and available moisture used. The irrigation treatments were planned to reflect the following relationships:

	Average percent of soil moisture	Average percent of available moisture used	Atmospheres of tension
I_1	8.6	94	12.0
I_2	9.8	82	5.0
I_3	11.6	63	1.0
I_4	13.1	47	0.82
I_5	14.4	34	0.66

Soil samples were also taken at harvest, and the amount of residual soil moisture was estimated. No significant amount of rainfall was recorded during the growing season. Soil at this site has an estimated available water-holding capacity of 7.3 inches in the upper 4 feet.

Treatment levels and corresponding yields are given in Table 6.13. Average yields per treatment combination are also listed. At fertilizer levels other than $N = 0$, the yield response to water is strong. The lack of response when $N = 0$ suggests a low level of preplant nitrogen in the soil. Average yields begin decreasing at a point beyond $N = 170$ pounds per acre when water treatments I_1, I_3, and I_5 are applied.

No block effect was observed for this experiment. The 1.5 polynomial in (6.24) was fitted to these data. Only those coefficients estimated for the nitrogen variables are statistically significant at conventional levels.

$$\text{yield} = 2923.2682 - 295.6678W_1 + 30.4974^{***}N + 46.8803W_1^{1.5}$$
$$\phantom{\text{yield} = }(5985.78) \qquad (571.99) \qquad (9.940) \qquad (67.72)$$

$$\phantom{\text{yield}}- 2.0224^{***}N^{1.5} + .2958W_1N \tag{6.24}$$
$$\phantom{\text{yield} - }(.351) \qquad\quad (.268)$$

$$R^2 = .818 \qquad F = 34.09^{***} \qquad LOFF = 2.872^{**}$$

The signs for the linear and 1.5 coefficients for the water variable would be expected to be plus or minus, respectively. They are reversed in (6.24). Finally, since the LOFF statistic is significant at the 5 percent level, the 1.5 polynomial does not adequately represent yield-water-nitrogen relationships for this experiment.

A plot of the $(Y - \hat{Y})$ residuals against W_1 shows that residuals corresponding to the observations generated by $W_1 = 24.0$, 37.7, and 40.6 are not well distributed about a zero line. When these observations are omitted, the 1.5 polynomial is reestimated as in (6.25). The LOFF statistic is considerably reduced. The coefficients for the water variables are still not statistically significant at conventional levels but their signs are correct. Based on Table 6.13, however, the average yield response to water is quite strong at all levels of fertilization other than $N = 0$.

$$\text{yield} = -17456.2964^{**} + 1071.8934W_1 + 41.5719^{***}N - 84.6556W_1^{1.5}$$
$$\phantom{\text{yield} = -}(8503.90) \qquad (700.72) \qquad (10.11) \qquad (78.41)$$

$$\phantom{\text{yield}}- 1.7044^{***}N^{1.5} - .1732W_1N \tag{6.25}$$
$$\phantom{\text{yield} - }(.394) \qquad\quad (.283)$$

$$R^2 = .840 \qquad F = 29.46^{***} \qquad LOFF = .847$$

When observed yields are plotted against the corresponding values for W_1, the yield response is nearly linear and (or) convex to the origin for $N \neq 0$. If the $W_1^{1.5}$ variable is eliminated, only a linear effect for water and an interaction with N are built into the model. The quantified version of this model is in (6.26) where all estimated coefficients except the interaction term are statistically significant at the 5 percent or lower level. In (6.24) and (6.25), the effect of water was being divided between W_1 and $W_1^{1.5}$, with the latter not being strongly reflected in plots of the data. In (6.26), the contribution of water has been collapsed into W_1 and W_1N with the coefficient for W_1 being highly significant.

$$\text{yield} = -8652.4044^{**} + 319.1605^{***}W_1 + 44.6535^{***}N - 1.6432^{***}N^{1.5}$$
$$\phantom{\text{yield} = -}(2419.59) \qquad (70.29) \qquad (9.726) \qquad (.391)$$

$$\phantom{\text{yield}}- .3085W_1N \tag{6.26}$$
$$\phantom{\text{yield} - }(.254)$$

$$R^2 = .834 \qquad F = 36.32^{***} \qquad LOFF = .905$$

Equation (6.25) is used for predicting yields at different levels of water and nitrogen. These yields are in Table 6.14.

Corn silage

The yield-water-nitrogen data for corn silage are also given in Table 6.13. Based on an analysis of variance, there was no block effect for corn silage. The square root function fitted to these data is in (6.27). Since the *LOFF* statistic is significant at the 5 percent level, the $(Y - \hat{Y})$ residuals were plotted against W_1 and N. The residuals generated by the $N = 255$ observations were not evenly distributed about a zero line. After omitting these four observations, the square root function is reestimated in (6.28). The *LOFF* for (6.28) is no longer statistically significant at the 5 percent level, and the equation can be used appropriately for estimating the yield-water-nitrogen relationships for silage. Predicted yields derived from (6.28) are in Table 6.15.

$$\text{yield} = -98703.3553 - 3418.2353W_1 - 58.3041^{***}N + 41177.2655W_i^{.5}$$
$$\quad (94494.14) \qquad (3125.67) \qquad (18.74) \qquad (34463.77)$$

$$- 4149.6734^{**}N^{.5} + 1049.1109^{***}W_i^{.5}N^{.5} \tag{6.27}$$
$$\quad (1760.88) \qquad (303.44)$$

$$R^2 = .910 \qquad F = 76.69^{***} \qquad LOFF = 2.596^{**}$$

$$\text{yield} = -168886.8208^{*} - 5762.8935^{*}W_1 - 59.7430^{***}N$$
$$\quad (94982.12) \qquad (3144.48) \qquad (17.94)$$

$$+ 66946.0624^{*}W_i^{.5} - 5152.0528^{***}N^{.5}$$
$$\quad (34662.14) \qquad (1745.60) \tag{6.28}$$

$$+ 1234.2855^{***}W_i^{.5}N^{.5}$$
$$\quad (301.96)$$

$$R^2 = .925 \qquad F = 83.60^{***} \qquad LOFF = 2.123$$

MESA, ARIZONA, 1970

Procedures and experimental design are similar to those used in the 1971 experiment. The seed variety planted was Funks G-711-AA. Prior to planting, 300 pounds of treble super phosphate per acre and 10 acre-inches of water were applied. Plants were thinned to approximate population of 20,000 plants per acre.

Irrigation scheduling was based on irrometer readings. Irrigation treatments reflected the following relationships:

	Average irrometer reading	Average percent of soil moisture	Atmospheres of tension
I_1	85+	10.0	5.0
I_2	75	11.2	1.5
I_3	59	12.4	0.9
I_4	45	13.4	0.7
I_5	31	14.5	0.6

Rainfall was not a factor, and five levels of fertilization were incorporated in the experiment. Abnormally high temperatures and low humidity at and shortly after pollination resulted in relatively low grain yields and high water requirements. Yields and levels of water and nitrogen application are listed in Table 6.16. A comparison of average yields with those in Table 6.13 shows average yields in 1970 were considerably below the 1971 averages.

Based on an analysis of variance, experimental units in Block 1 were substantially different from those in Block 2 so that a block effect resulted. The 1.5 polynomial in (6.29) represents yield-water-nitrogen relationships for this experiment. Variable X_1 is a dummy variable used as a proxy for the block effect. All estimated coefficients are statistically significant at the 10 percent or lower level.

$$\text{yield} = -6367.7492^{**} - 264.0144^{**}X_1 + 502.6979^{***}W_1 + 7.2824^{*}N$$
$$\phantom{\text{yield} = }(2849.52) \quad\;\; (127.37) \qquad\;\; (184.73) \qquad (4.025)$$

$$- 49.5216^{***}W_1^{1.5} - .7347^{***}N^{1.5} + .2094^{***}W_1 N$$
$$\;\;(17.94) \qquad\quad (.167) \qquad\;\; (.055)$$

$$R^2 = .777 \qquad F = 21.54^{***} \qquad LOFF: \text{Block 1} = 1.643$$
$$\text{Block 2} = 1.023$$

(6.29)

When the water variable is defined in terms of water used, W_2, the square root function in (6.30) provides a good fit to the observed data.

$$\text{yield} = -21585.5203^{**} - 261.6818^{**}X_1 - 536.9145^{**}W_2 - 4.8191^{**}N$$
$$\phantom{\text{yield} = }(9251.62) \qquad\;\; (118.25) \qquad\;\; (205.62) \qquad (1.860)$$

$$+ 7006.7411^{**}W_2^{.5} - 243.6380^{**}N^{.5}$$
$$\;\;(2766.32) \qquad\quad (94.91)$$

$$+ 60.0235^{***}W_2^{.5}N^{.5}$$
$$\;\;(13.13)$$

$$R^2 = .808 \qquad F = 25.99^{***} \qquad LOFF: \text{Block 1} = 1.056$$
$$\text{Block 2} = 0.736$$

(6.30)

All estimated coefficients are statistically significant at the 5 percent or lower level. The coefficient for the $N^{.5}$ variable is negative; a positive coefficient would be expected. Using (6.29) as a predicting equation, derived yields for several combinations of water and nitrogen are summarized in Table 6.17.

Corn silage

Water and nitrogen treatments and corresponding yields are given in Table 6.16. No block effect was determinable. Considering the 1.5 polynomial in (6.31), all estimated coefficients other than for the intercept are statistically significant at the 5 percent or lower level.

$$
\begin{aligned}
\text{yield} = &-55265.6614 + 5331.3322^{**}W_1 + 156.1282^{***}N - 514.8577^{**}W_1^{1.5} \\
&\;\;(34447.35) \qquad (2233.64) \qquad (48.52) \qquad\quad (216.89) \\
&- 9.4535^{***}N^{1.5} + 1.3959^{**}W_1 N \\
&\;\;\;(2.020) \qquad\;\; (.658)
\end{aligned}
\tag{6.31}
$$

$$
R^2 = .758 \qquad F = 23.82^{***} \qquad LOFF = 1.155
$$

Predicted yields estimated from (6.31) are listed in Table 6.18.

YUMA MESA, ARIZONA, 1970

Funks G-4384 was planted in 40-inch rows on Superstition Fine Sand. Plant population was estimated as 19,600 per acre. Several irrigations were necessary for germination and establishment of stand on this sandy soil. Irrigation treatments $I_1, \ldots, I_5$ were made when the percent soil moisture dropped to 4.5, 5.1, 5.6, 6.2, and 6.7 percent, respectively. Because of the low available water-holding capacity of this soil, 3.1 acre-inches in the upper 4 feet, irrigation was frequent. The number of irrigations over the growing season ranged from 10 to 16. No significant amount of rainfall was recorded.

Prior to planting, 100 pounds of P_2O_5 per acre and 25 pounds of nitrogen were uniformly applied to the experimental area. Since soil nutrients are easily leached, the level of preplant nitrogen in the soil was assumed to be zero. All nitrogen was added to water. The rate of application depended on the total quantity of nitrogen to be applied and the anticipated number of irrigation applications. Levels of water and nitrogen applications and corresponding yields are summarized in Table 6.19.

No block effect among the three blocks was observed. The *LOFF* statistic for the quadratic function in (6.32) is very high. When the $(Y - \hat{Y})$ residuals were plotted against values for W and N, the residuals associated with the observations generated by $N = 150$ and $N = 300$ were not well distributed about a zero line. When these observations are omitted, the quadratic function is reestimated, as in (6.33). The improvement over (6.32) is not considerable. Referring to Table 6.19, the variation among observed yields for most treatment combinations is not high.

$$\text{yield} = -104357.8048^{***} + 6754.8649^{***}W - 37.5104^{***}N - 105.0133^{***}W^2$$
$$\phantom{\text{yield} = } (14200.23) \qquad (897.09) \qquad (9.382) \qquad (14.13)$$

$$- .0068N^2 + 1.2414^{***}WN \qquad\qquad\qquad (6.32)$$
$$ (.005) \qquad (.279)$$

$$R^2 = .581 \qquad F = 16.65^{***} \qquad LOFF = 7.804^{***}$$

Consequently, the pure error sums of squares for this experiment tends to be low and the *LOFF* statistic tends to be high.

$$\text{yield} = -92788.0699^{***} + 5978.6440^{***}W - 29.0719^{***}N - 92.2975^{***}W^2$$
$$\phantom{\text{yield} = } (18153.30) \qquad (1143.25) \qquad (8.059) \qquad (17.89)$$

$$- .0137^{***}N^2 + 1.0798^{***}WN \qquad\qquad\qquad (6.33)$$
$$ (.005) \qquad (.239)$$

$$R^2 = .710 \qquad F = 23.54^{***} \qquad LOFF = 6.288^{***}$$

Predicted yields derived from (6.33) are in Table 6.20. When water is 26 acre-inches or less or 41 acre-inches or greater, predicted yields are negative. The levels of water treatment for this experiment ranged only from 29.1 to 36.9 acre-inches.

Corn silage

No block effect was estimated among the corn silage yields. The quadratic function in (6.34) was fitted to these data. Based on plots of residuals, the residuals associated with $W = 31.4$ and $N = 300$ were not evenly distributed about a zero line. Deletion of these observations, however, did not result in a model superior to (6.34). Predicted yields derived from (6.34) are given in Table 6.21.

$$\text{yield} = -430326.0902^{***} + 28404.0246^{***}W_1 - 253.6018^{***}N - 441.0659^{***}W_1^2$$
$$\phantom{\text{yield} = } (115383.82) \qquad (7295.93) \qquad (75.75) \qquad (114.82)$$

$$- .0343N^2 + 8.2875^{***}W_1N \qquad\qquad\qquad (6.34)$$
$$ (.041) \qquad (2.257)$$

$$R^2 = .458 \qquad F = 10.14^{***} \qquad LOFF = 3.336^{***}$$

YUMA VALLEY, ARIZONA, 1970

Funks G-4384 was planted in 40-inch rows on Glendale Silty Clay Loam at the Yuma Valley site. Plant population approximated 19,600 plants per acre. One hundred pounds of P_2O_5 per acre were applied prior to planting.

Irrigation treatments $I_1, \ldots, I_5$ were made when the percent of available soil moisture was depleted to an average of 85, 75, 65, 55, and 45 percent, respectively. The number of irrigations ranged from three to six. This number is considerably below that for the Yuma Mesa experiment. The sites are only about 10 miles apart, but with considerably different soil characteristics. The estimated available water-holding capacity of soil at Yuma Valley is 9.9 acre-inches in the upper 4 feet compared with only 3.1 acre-inches at the Yuma Mesa site. Rainfall was not a factor in this experiment.

Yields and levels of water and nitrogen applications are summarized in Table 6.22. There was no block effect for this experiment. Among the commonly used polynomials, the square root function in (6.35) seemed appropriate. The *LOFF* statistic, however, is highly significant.

$$\text{yield} = 33044.412\overset{**}{1} + 1361.609\overset{**}{4}W_1 - 8.474\overset{**}{2}N - 12357.400\overset{**}{7}W_i^{.5}$$
$$\quad\ \ (13398.36)\qquad (559.19)\qquad (3.295)\qquad (5509.04)$$

$$+ 168.1146N^{.5} + 14.2498W_i^{.5}N^{.5} \tag{6.35}$$
$$\quad\ (162.31)\qquad\quad (31.10)$$

$$R^2 = .589 \qquad F = 17.\overset{**}{2}\overset{*}{1} \qquad LOFF = 5.\overset{***}{2}96$$

In plots of the $(Y - \hat{Y})$ residuals against W_1 and N, the residuals associated with the observations generated by $W = 24.5$ and 29.0 were not well distributed. When these 12 observations are omitted, the 1.5 polynomial in (6.36) is estimated. The *LOFF* statistics dropped considerably to 2.66, which is still statistically significant at the 5 percent level but is close to the tabled value of 2.45. The estimated standard errors of coefficients for the water variables in (6.36) are very high.

$$\text{yield} = 2566.2928 + 145.1635W_1 + 25.97\overset{***}{3}9N - 3.0827W_1^{1.5}$$
$$\quad\ \ (4519.09)\qquad (572.85)\qquad (6.579)\qquad (77.05)$$

$$- .91\overset{***}{7}8N^{1.5} - .1252W_1N \tag{6.36}$$
$$\quad\ (.325)\qquad\quad (.238)$$

$$R^2 = .692 \qquad F = 21.\overset{***}{5}9 \qquad LOFF = 2.6\overset{**}{6}0$$

A plot of observed yields against W_1 shows that the yield response to water is nearly linear and (or) convex to the origin for most levels of nitrogen. The yield response curve to water when $N = 160$ is concave to the origin. Based on these relationships, the contribution of water to yield is being divided between W_1 and $W_1^{1.5}$ even though the latter effect is not strongly present in these yield response curves. When the $W_1^{1.5}$ variable is suppressed, the modified 1.5 polynomial in (6.37) is derived.

$$\text{yield} = 2744.22\overset{***}{3}0 + 122.27\overset{***}{5}4W_1 + 25.99\overset{***}{8}4N - .91\overset{***}{6}7N^{1.5}$$
$$\quad\ \ (792.91)\qquad (28.84)\qquad (6.484)\qquad (.321)$$

$$- .1265W_1N \tag{6.37}$$
$$\quad\ (.233)$$

$$R^2 = .692 \qquad F = 27.\overset{***}{5}5 \qquad LOFF = 2.217$$

All coefficients except for the interaction term are now highly significant. The *LOFF* is no longer significant. Thus the model in (6.37) is plausible for quantifying yield-water-nitrogen relationships when the 12 observations corresponding to $W = 24.5$ and 29.0 are omitted. Predicted yields derived from (6.37) are in Table 6.23.

Corn silage

Yields and treatment levels are given in Table 6.22. No block effect was determined for these data. The 1.5 polynomial in (6.38) was selected to represent this experiment. Only those coefficients estimated for the nitrogen variable are statistically significant at conventional levels. Predicted yields are listed in Table 6.24.

$$\text{yield} = 38918.1169 - 1955.5261W_1 + 188.7313\overset{***}{N} + 342.9222W_1^{1.5}$$
$$\quad (40656.49) \quad (5277.94) \quad\quad (63.82) \quad\quad (716.25)$$

$$\quad - 10.4150\overset{***}{N}^{1.5} + .7974W_1N \quad\quad\quad\quad\quad (6.38)$$
$$\quad\quad (2.504) \quad\quad\quad (1.945)$$

$$R^2 = .511 \quad\quad F = 7.94\overset{***}{} \quad\quad LOFF = .720$$

SAFFORD, ARIZONA, 1972

Texas 28A was planted in 40-inch rows on Pima Clay Loam Variant. The estimated plant population was 13,100 per acre. Irrigation treatments $I_1, \ldots, I_5$ were made when the percent of available moisture used dropped to an average of 85, 70, 55, 40, and 25 percent, respectively. The nitrate content of the water applied at this site is relatively high with a nitrogen equivalent of 4 pounds per acre-inch of water. Rainfall exceeding 0.25 inch at any occurrence totaled about 8 inches during the growing season. The estimated available water-holding capacity of the soil at this site is 7.5 acre-inches in the upper 4 feet.

Yield and application levels of water and nitrogen are summarized in Table 6.25.

There was no block effect for this experiment. No strong relationship between yields and water and (or) nitrogen was observed. The quadratic function fitted to the data is given in (6.39). None of the estimated coefficients were statistically significant at conventional levels.

$$\text{yield} = -9032.1329 + 320.2190W_1 - 4.2171N - 2.3249W_1^2$$
$$\quad (13474.70) \quad\quad (470.96) \quad\quad (8.516) \quad\quad (4.101)$$

$$\quad .0015N^2 + .0987W_1N \quad\quad\quad\quad\quad (6.39)$$
$$\quad (.007) \quad\quad (.144)$$

$$R^2 = .346 \quad\quad F = 3.918\overset{***}{} \quad\quad LOFF = 1.232$$

When plotted against W_1 and N the $(Y - \hat{Y})$ residuals were reasonably well-distributed about a zero line. Predicted yields derived from (6.39) are in Table 6.26.

PLAINVIEW, TEXAS, 1971

Pioneer 3306 was planted in 40-inch rows on Pullman Clay Loam at the High Plains Research Foundation, Plainview, Texas. Plant population approximated 22,400 plants per acre.

The irrigation treatments were applied according to the stage of plant growth. The irrigation scheduling program and quantities applied are identified in Table 6.28. Rainfall exceeding 0.25 inch at any occurrence totaled 10.7 inches during the growing season. Soil at this site has an estimated available water-holding capacity of 8 acre-inches in the upper 4 feet. An estimated 18 pounds of nitrogen per acre in nitrate form were available prior to any fertilizer applications. One hundred pounds of phosphate per acre in the form of 0-46-0 were applied prior to planting.

Treatment levels and corresponding yields are summarized in Table 6.27. Since each treatment combination is fully replicated, the experimental design used in this test is a "randomized complete block with factorials." As noted earlier, Table 6.28 contains irrigation dates and quantities applied. This table is of interest in interpreting the average yield response to water at different levels of fertilization. Irrigation treatments I_1, I_2, and I_3 are identical in terms of quantity applied and dates of application. At any level of nitrogen, however, individual and average yields among the three irrigation treatments are quite variable. Plant population differed slightly among some of the three treatments. These population differences, however, were less than 1000 plants per acre. Other than plant population, differences in individual and average yields generated by I_1, I_2, and I_3 are due to variability in soil, management, and unexplainable random occurrences.

The quantity of water applied with I_4 and I_5 is essentially identical. An important difference in irrigation scheduling is indicated in Table 6.28. With I_4, 13.9 acre-inches were applied on June 25 while for I_5 13.3 acre-inches were applied on July 14. The latter application was only 0.6 acre-inch less but was applied nearly three weeks later than the I_4 treatment. This difference in timing apparently had an important impact on yield. Average yields associated with I_5 are less than half those generated by I_4 at all levels of fertilization.

The impact of varying plant population on observed yields was not statistically significant. The 1.5 polynomial in (6.40) represents yield-water-nitrogen relationships for this experiment. All estimated coefficients other than for N_1 were highly significant. Based on the low value for the *LOFF* statistic, the 1.5 polynomial appears adequate for quantifying the data generated by this experiment. Predicted yields derived from (6.40) are given in Table 6.29.

$$\text{yield} = -769863.9795^{***} + 49557.1052^{***}W_1 + 9.8801N_1$$
$$(65014.37) \qquad (4188.59) \qquad (6.165)$$
$$- 4783.3357^{***}W_1^{1.5} - 1.4592^{***}N_1^{1.5} + .3927^{***}W_1N_1$$
$$(405.42) \qquad (.240) \qquad (.086) \qquad \textbf{(6.40)}$$

$$R^2 = .883 \qquad F = 142.14^{***} \qquad LOFF = 0.515$$

PLAINVIEW, TEXAS, 1970

Pioneer 3306 was planted in 40-inch rows at the same site as the 1971 test. The planned plant population was 24,000 plants per acre. As with the 1971 experiment, irrigation treatments were applied according to stage of plant growth. Dates of application and quantities applied are listed in Table 6.31. Rainfall exceeding 0.25 inch at any occurrence totaled 4.8 inches during the growing season. Nitrogen treatments were identical to those for the 1971 experiment. Based on analysis of soil samples, an average of 39 pounds of nitrogen per acre in the form of nitrate were available prior to any fertilizer applied in 1970.

Levels of water and nitrogen applications, plant population, and corresponding yields are summarized in Table 6.30. The average yields associated with I_2 and I_3 are of interest. Average yields generated by I_3 are more than double these of I_2. The difference in total water applied, however, is only 2.1 acre-inches. The considerably higher yields with I_3 would appear to result from a difference in irrigation scheduling. The difference is specified in Table 6.31 where 9.8 acre-inches were applied for I_3 at the tasseling stage while for I_2 7.7 acre-inches were divided between applications seven weeks after germination and silking.

The yield response to fertilizer does not exhibit any strong, consistent pattern. A yield response curve concave to the origin would be expected. When the average yields are plotted, however, the average yield response curves are of the sine-cosine form at all levels of water applied. That is, average yields do not exhibit any consistent trend. This phenomenon is reflected in the 1.5 polynomial in (6.41) fitted to these data. Both coefficients for the nitrogen variables are negative and neither is statistically significant at conventional levels. Furthermore, the $LOFF$ statistic is highly significant.

$$\text{yield} = -27169.9699^{***} + 3178.0990^{***}W_1 - 1.8243N_1 - 381.0653^{***}W_1^{1.5}$$
$$\qquad (3133.41) \qquad (333.20) \qquad (8.443) \qquad (41.69)$$

$$\qquad - .1288N_1^{1.5} + .1903W_1N_1 \qquad\qquad\qquad\qquad (6.41)$$
$$\qquad (.373) \qquad\quad (.148)$$

$$R^2 = .706 \qquad F = 44.59^{***} \qquad LOFF = 3.232^{***}$$

When the $(Y - \hat{Y})$ residuals are plotted against W_1 and N_1, the residuals corresponding to the observations generated by I_3 or $W_1 = 16.6 + 4.8$ are not evenly distributed above a zero line. After omitting these observations, the 1.5 polynomial is reestimated as in (6.42). The new $LOFF = 0.428$ is not significant, and (6.42) is plausible or appropriate for quantifying yield-water-nitrogen relationships when the I_3 observations are omitted. Predicted yields derived from (6.42) are listed in Table 6.32.

$$\text{yield} = -30648.0274^{***} + 3427.2477^{***}W_1 - 2.0606N_1 - 405.6546^{***}W_1^{1.5}$$
$$\quad\quad (2554.60) \quad\quad (267.47) \quad\quad (7.549) \quad\quad (33.37)$$

$$- .0098N_1^{1.5} + .1274W_1N_1 \quad\quad\quad\quad\quad\quad\quad\quad (6.42)$$
$$\quad (.332) \quad\quad (.127)$$

$$R^2 = .848 \quad\quad F = 81.34^{***} \quad\quad LOFF = .428$$

PLAINVIEW, TEXAS, 1969

Pioneer 3306 was planted in 40-inch rows and plant population was an approximated 26,000 plants per acre. Irrigation treatments were applied according to the stage of plant growth. The irrigation scheduling program and quantities applied are identified in Table 6.34. Rainfall exceeding 0.25 inch at any occurrence totaled 16.3 inches during the growing season. Nitrogen treatments are identical to those for the 1970 and 1971 trials. The level of preplant nitrogen in the form of nitrate was estimated at 16 pounds per acre.

Treatment levels and corresponding yields are in Table 6.33. Several relationships between Tables 6.33 and 6.34 are of interest. A few examples will be illustrated. When $N = 0$, the average yield generated by $I_5 = 28.8$ is about 920 pounds below the average yield of 5928 pounds associated with $I_4 = 22.6$ inches. The only difference between I_4 and I_5 indicated in Table 6.34 is that with I_5 an additional 6.2 acre-inches were applied during the soft dough stage. The yields generated by I_5 are lower than those for I_4 at all levels of fertilization even though I_5 involves an additional 6.2 acre-inches of water. In fact, the 28.8 acre-inches applied for I_5 are higher than the quantities applied for I_2, I_3, and I_4. Average yields generated by I_5, however, are less at all levels of nitrogen. This relationship indicates that in this test the timing of application is more important than the quantities applied. The same conclusion holds when average yields associated with I_3 are compared with those for I_4. The average yield differential is not substantial until $N_5 = 300$ pounds per acre is applied. I_4, however, involves an additional application of 5.2 acre-inches of water at tassel initiation.

One final comparison is made. When $N = 200$ or higher, the average yields generated by I_1 and I_2 are consistently and substantially above those for the other irrigation treatments. I_2 represents the second lowest quantity of water applied. The important feature shared only by I_1 and I_2 is that 5.5 acre-inches of water were applied during the silking period. This irrigation appears to have had a strategic effect on yield when relatively heavy quantities of nitrogen were applied.

The quadratic function in (6.43) represents yield-water-nitrogen relationships for this experiment. The *LOFF* statistic is highly significant. When the $(Y - \hat{Y})$ residuals were plotted against W_1 and N_1, residuals associated with I_2 and I_5 were not well distributed about a zero line. After deleting these observations, the quadratic is reestimated, as in (6.44). Predicted yields estimated from (6.44) are given in Table 6.35.

$$\text{yield} = 28925.9397^{***} - 1116.3451^{***}W_1 + 12.4363^{**}N_1$$
$$(4226.95) \qquad (188.70) \qquad (6.296)$$

$$+ 13.0438^{***}W_1^2 - .0168N_1^2 - .0543W_1N_1 \qquad \qquad \textbf{(6.43)}$$
$$(2.074) \qquad (.011) \qquad (.120)$$

$$R^2 = .469 \qquad F = 16.44^{***} \qquad LOFF = 2.836^{***}$$

$$\text{yield} = 19734.7845^{***} - 685.5595^{***}W_1 + 6.7974N_1$$
$$(5521.87) \qquad (252.34) \qquad (6.158)$$

$$+ 8.2994^{***}W_1^2 - .0154N_1^2 + .0424W_1N_1 \qquad \qquad \textbf{(6.44)}$$
$$(2.777) \qquad (.012) \qquad (.108)$$

$$R^2 = .537 \qquad F = 12.52^{***} \qquad LOFF = 1.110$$

PLAINVIEW (LAKE SITE), TEXAS, 1971

Pioneer 3306 was planted in 40-inch rows on Randall Clay at Plainview. The experimental site is a former lake bed. The planned plant population was 22,400 plants per acre but actual population was considerably less.

The irrigation scheduling program and quantities applied are listed in Table 6.37. Irrigation treatments I_1, I_2, and I_3 are identical. Rainfall for this experiment totaled 9.6 inches. The Randall Clay soil has an estimated available water-holding capacity of 8.0 acre-inches in the top 4 feet.

The fertilizer treatments are the same as for the 1969–71 experiments at Plainview. Based on soil samples, the level of preplant nitrogen was estimated at 36 pounds per acre.

Treatment levels, average plant population, and associated yields are given in Table 6.36. As noted earlier, treatments I_1, I_2, and I_3 are identical. Average yields at lower levels of fertilization, however, are quite variable. The range of quantities applied for $I_1, \ldots, I_5$ is narrow. Differences in timing of application, however, appear to have an important impact on subsequent yields. The difference in quantity applied between I_1–I_3 and I_5 is only 1.5 acre-inch, but the timing of irrigation outlined in Table 6.37 is substantially different. As a consequence, average yields generated by I_5 at levels of nitrogen other than $N = 0$ are considerably below those for I_1–I_3.

Plant population had a significant effect on yield. The square root function fitted to the data is in (6.45) where P is plant population in terms of 10,000 plants per acre. Coefficients for the intercept and water variables are statistically significant at the 11 percent level. The magnitudes of several coefficients are unusually high. Since plant population varies with experimental plot, the $LOFF$ statistic cannot be determined for this experiment. Both coefficients for the nitrogen variables have negative values; the signs for the water variables coefficients are opposite from expected values.

$$\text{yield} = 961093.2465 + 1265.9910\overset{**}{P} + 19450.0839W_1 - 12.9884\overset{**}{N_1}$$
$$(592400.74) \quad (599.74) \quad (12000.60) \quad (5.793)$$

$$- 273498.8682W_i^{.5} - 2213.1117\overset{**}{N_i^{.5}}$$
$$(168641.62) \quad (1267.13) \tag{6.45}$$

$$+ 363.0381\overset{**}{W_i^{.5}N_i^{.5}}$$

$$R^2 = .395 \quad F = 10.10\overset{***}{}$$

Plots of the $(Y - \hat{Y})$ residuals against W_1 and N_1 are evenly distributed. The average yield response curves to water are linear or convex to the origin for the various levels of nitrogen. The conclusion is that the range of water treatments is too narrow. The average yield response curves to nitrogen on the other hand are of variable shape. Average yield for I_4 first decreases then increases and finally decreases again. A similar pattern is evident for I_3. There is no biological basis for this pattern of yields. Since the square root effect for water does not appear to exist and that for nitrogen is indeterminate, the square root model was refitted to the data but with these variables deleted. All estimated coefficients in (6.46) are statistically significant at the 5 percent or lower level. Predicted yields derived from (6.45) are given in Table 6.38.

$$\text{yield} = -14599.1672\overset{***}{} + 1258.4319\overset{**}{P} + 286.6090\overset{***}{W_1} - 14.0734\overset{**}{N_1}$$
$$(3011.05) \quad (610.07) \quad (60.46) \quad (5.871)$$

$$+ 54.4711\overset{**}{W_i^{.5}N_i^{.5}} \tag{6.46}$$
$$(20.19)$$

$$R^2 = .358 \quad F = 13.25\overset{***}{}$$

ANALYSIS OF WHEAT EXPERIMENTS

The geographical coverage of wheat experiments was not as extensive as for corn. Experiments were conducted only in Colorado and Arizona but at four different sites in Arizona. The introductory discussion for Chapter 6 also applies here and the format for this chapter is the same. The 1971–72 experiment at Yuma Valley, Arizona, is analyzed in detail.

YUMA VALLEY, ARIZONA, 1971–72

Wheat was planted on Glendale Silty Clay Loam at the Yuma Valley experimental site by drilling Inia 66 at a rate of 75 pounds per acre. A preplant irrigation of 9.6 acre-inches raised the soil moisture level to field capacity. This soil has an estimated available water-holding capacity of 9.9 inches in the upper 4 feet. Irrigation treatments $I_1, \ldots, I_5$ were made when the percentage of available water used in the second foot of soil dropped to an average of 85, 75, 65, 55, and 45 percent, respectively.

Estimates of preplant soil nitrogen averaged 51 pounds per acre across the experimental site. Fertilizer treatments $N = 50$ and 100 were applied with the first irrigation. For the $N = 150$ and 200 treatments, 100 pounds were applied with the first irrigation and the rest with an irrigation when plants were at or close to the jointing stage. No appreciable amount of rainfall occurred during the growing season. Treatment levels and corresponding yields are summarized in Table 7.1.

No block effect among the three blocks each containing 22 plots was observed. The quadratic function in (7.1) represents the yield-water-nitrogen relationships selected statistically for prediction. Based on the *LOFF* statistic, the quadratic model is adequate for representing these data. Coefficients for the treatment variables have the expected signs; the levels of statistical significance are indicated where * = 0.10, ** = 0.05, and *** = 0.01 levels.

$$\text{yield} = -10414.9628^{**} + 852.0111W_1 + 11.6046^{*}N_1 - 12.9168^{*}W_1^2$$
$$(5199.42) \qquad (396.56) \qquad (6.472) \qquad (7.537)$$

$$- .0320^{**}N_1^2 + .0925W_1N_1 \qquad\qquad (7.1)$$
$$(.013) \qquad (.219)$$

$$R^2 = .762 \qquad F = 38.35^{***} \qquad LOFF = 1.273$$

Measurements of soil moisture at harvest are also available. The quadratic formulation in terms of W_2 is in (7.2). None of the coefficients are statistically significant at the 10 percent or lower level. In addition, the signs for the linear and quadratic effects of W_2 are opposite of those expected.

$$\text{yield} = 1414.1452 - 35.4620W_2 + 7.8317N_1 + 5.6855W_2^2 - .0185N_1^2$$
$$\quad\quad (4415.01) \quad\quad (423.73) \quad\quad (7.754) \quad\quad (10.10) \quad\quad (.014)$$

$$-.0733W_2N_1 \quad\quad\quad\quad\quad\quad\quad\quad\quad\quad\quad (7.2)$$
$$\quad (.343)$$

$$R^2 = .764 \quad\quad F = 38.80^{***} \quad\quad LOFF = 1.194$$

A five-parameter Mitscherlich function of the form in (3.8) was also fitted to the 66 observations. Available soil moisture prior to first irrigation treatment (w) and residual or preplant nitrogen (n) were estimated as 9.9 acre-inches and 51 pounds per acre, respectively. The quantified form is in (7.3) where $W' = W$ applied $+ 9.9$ and $N' = N$ applied $+ 51$.

$$\text{yield} = 5566.23^{***} \left[1 - e^{(.0683)} e^{\frac{.0657W' - .0049W'^2}{(.0034)}} \right]$$
$$\quad (1200.13)$$

$$\left[1 - e^{\frac{-.0348N'}{(.0037)}} e^{\frac{.000068N'^2}{(.00023)}} \right] \quad\quad (7.3)$$

$$R^2 = .765 \quad\quad LOFF = 1.020$$

The estimated maximum yield is 5566 pounds per acre. In comparison, the maximum yield derived from (7.1) is 5322 pounds per acre. Only the coefficients for maximum yield and N' in (7.3) are statistically significant at conventional levels. The signs for the W' and W'^2 are reversed. A negative coefficient for W', which appears positive when substituted into the Mitscherlich formulation, implies that the "growth factor" for water and (or) the efficiency with which a plant uses available water is negative. This is not consistent with the yield response to water when nitrogen is held constant at the specified levels in Table 7.1. The signs for the coefficients estimated for the N' variables are consistent with hypothesized relationships. Even though some similar properties exist between (7.1) and (7.3), the estimating procedure for deriving (7.3) is relatively complex. Predicted yields derived from (7.1) are given in Table 7.2.

A three-dimensional production surface based on points estimated from (7.1) appears in Figure 7.1. Curve AB, for example, represents the yield response to increased applications of water when N is fixed at 20 pounds per acre. Yield begins decreasing at some point beyond $W = 34$, and the MP_W becomes negative.

Yield response curves to successively larger applications of water at three levels of fertilizer are plotted in Figure 7.2. This figure simply represents a plotting of points abstracted from Table 7.2 and implicit in Figure 7.1. Recall that about 50 pounds of residual soil nitrogen were available at

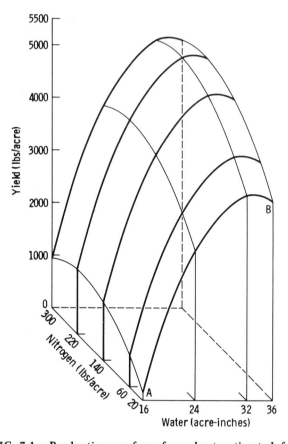

FIG. 7.1. Production surface for wheat estimated from
(7.1) according to specified water and nitrogen
levels (Yuma Valley, Arizona, 1971–72).

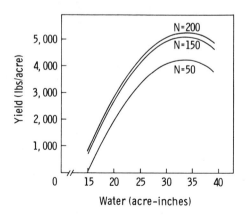

FIG. 7.2. Wheat yield response curves to water derived
from (7.1) assuming specified levels of fertilizer
(Yuma Valley, Arizona, 1971–72).

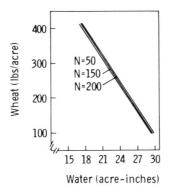

FIG. 7.3. Marginal product curves for water in production
of wheat at specified levels of nitrogen (Yuma
Valley, Arizona, 1971-72).

planting. Curve $N = 50$ in Figure 7.2 thus reflects yield response to water
when no nitrogen application is made. Yield reaches a maximum in the
vicinity of $W = 33$. When 100 pounds of N per acre are applied to the 50
already in the soil, the yield response curve substantially shifts upward but
still reaches a maximum when $W \simeq 33$. The contribution to yield of an additional 50 pounds of N, that is, $N = 200$, is relatively modest.

Marginal product curves for water and nitrogen are plotted in Figures
7.3 and 7.4, respectively. These curves are based on relationships implicit in
(7.1). The equation for the marginal product of water is given in (7.4) where
the subscripts are omitted. Equation (7.4) is linear; the marginal product of

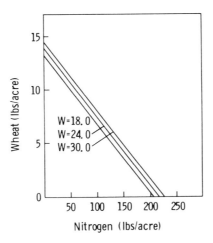

FIG. 7.4. Marginal product curves for nitrogen in production of wheat at specified levels of water (Yuma
Valley, Arizona, 1971-72).

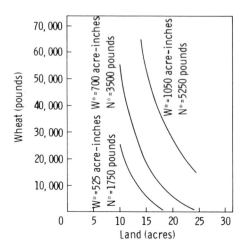

FIG. 7.5. Marginal product curves for land in production
of wheat with specified availabilities of water
and nitrogen (Yuma Valley, Arizona, 1971-72).

water, MP_W or $\partial Y/\partial W$, decreases by 25.83 pounds for each successive acre-inch of water at all levels of fertilization. Selecting any level of W, the MP_W or $\partial Y/\partial W$ increases by only 9.25 and 4.63 pounds when N is increased from 50 to 150 and 200 pounds per acre, respectively. The marginal product curves for nitrogen are also linear. This is a mathematical property of a quadratic function.

$$\partial(\text{yield})/\partial W = MP_W = 852.0111 - 25.8336W + .0925N \qquad (7.4)$$

The equation for the marginal product of land is in (7.5) where A = number of acres, yield* = yield $\times A$, $W^* = W \times A$, and $N^* = N \times A$. Equation (7.5) is derived by the same procedure outlined for (6.9).

$$\partial(\text{yield}^*)/\partial A = -10414.9628$$
$$+ [12.9168(W^*)^2 + .0320(N^*)^2 - .0925W^*N^*]/A^2 \qquad (7.5)$$

Three levels of W^* and N^* for application on successively larger areas of land have been assumed. The MP_A curves in Figure 7.5 are curvilinear because of the A^2 term in (7.5). The marginal product of land declines rapidly. When, for example, $W^* = 700$ acre-inches and $N^* = 3500$ pounds are applied on 10, then 12, and finally 14 acres, the marginal product of land, MP_A or $\partial(\text{yield})/\partial A$, declines from about 54,530 to 34,690 to 22,720 pounds, respectively. The MP_A, however, increases considerably when larger amounts of W^* and N^* are available for application to a fixed number of acres. When $A = 15$, for example, the increase in MP_A resulting from higher values for W^* and N^* is represented by the vertical distance between the curves.

The isoquant equation for (7.1) is given in (7.6). As an alternative, the

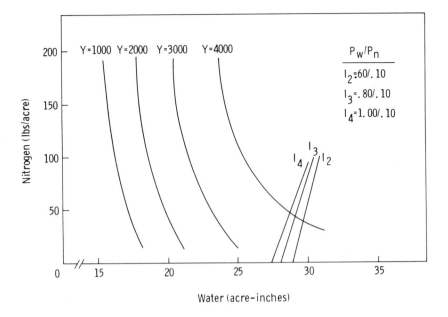

FIG. 7.6. Isoquants and isoclines for per acre yield of wheat at specified price ratios (Yuma Valley, Arizona, 1971-72).

equation could have been written in terms of N. The isoquants in Figure 7.6 are plotted by specifying yield = 1000, for example, assuming different levels of N and then determining the corresponding levels of W. Marginal rates of substitution (MRS) between W and N are derived by taking the partial derivative of (7.6) with respect to N.

$$W = \{852.0111 + .0925N \pm [(-852.0111 - .0925N)^2 - (51.6672) \text{ (yield}$$

$$+ 10414.9628 - 11.6046N + .0320N^2)]^{.5} \}/.0640 \qquad (7.6)$$

The value of MRS at a point is estimated by substituting the values for yield, W and N corresponding to that point on the isoquant.

The shapes of the isoquants in Figure 7.6 indicate a very high MRS of W for N. Recall that the slope of a line tangent to any point on an isoquant equals the MRS between the two inputs at that point. MRS also represents the ratio of the marginal products of the inputs. As the level of output increases, the isoquants have more curvature and, therefore, a greater range of substitutability between water and nitrogen. The least-cost combination of W and N is where the MRS of W for N equals P_w/P_n. Using the price ratios specified in Figure 7.6, only isoquant $y = 4000$ has sufficient curvature so that both W and N are included in the least-cost input mix. At output levels less than $y = 4000$, $MRS = MP_W/MP_N > P_w/P_n$ or $MP_W/P_w > MP_N/P_n$ and production should be generated from water applied to existing nutrients in the soil. Under these price and input-output relationships, application of nitrogen is not profitable. As P_w becomes increasingly expensive relative to P_n, a point is reached where nitrogen can profitably be substituted for water. With isoquants $y = 1000$ and $y = 2000$, this price ratio must be extremely high before movement away from the water axis and along an isoquant occurs.

YUMA VALLEY, ARIZONA, 1970–71

Seed and seeding rates were unchanged from the 1971–72 experiment. The experimental plots received 100 pounds of P_2O_5 per acre during seedbed preparation. A germination irrigation of 9.6 inches was applied to bring the soil moisture level in the top 4 feet to field capacity. Irrigation scheduling was the same as for the 1971–72 experiment. No significant amount of rainfall occurred during the growing season. Fertilizer treatments $N_2 = 75$ and $N_3 = 150$ were applied prior to planting. The $N_4 = 225$ and $N_5 = 300$ treatments were split. One-half was applied prior to planting; the rest was applied with a scheduled irrigation before the plants reached the boot stage.

Treatment levels and corresponding yields are given in Table 7.3. Water treatments I_1 and I_3 were repeated at the higher nitrogen levels. The other irrigation treatments vary with the nitrogen treatment. When the water treatment is relatively low at I_1 and I_2, nitrogen applications of $N = 300$ and $N = 225$, respectively, are too heavy and average yields decrease. For I_3, I_4, and I_5, average yield per acre-inch of water is increasing as larger amounts of nitrogen are applied.

No block effect was observed for this experiment. Coefficients estimated for a 1.5 polynomial are in (7.7). The *LOFF* statistic is extremely high. The lack of statistical significance for the W_1 coefficients is surprising. Referring to Table 7.3 and choosing any nitrogen level, average yields are increasing as additional water is applied. For $N = 150$, however, average yields drop but then increase substantially. When the $(Y - \hat{Y})$ residuals are plotted against W_1 and N, the residuals corresponding to the observations for $N = 75$ and $N = 225$ are not evenly distributed about a zero line. When these eight observations are omitted and the 1.5 polynomial refitted, the *LOFF* is essentially the same as in (7.7). Returning to Table 7.3, average yields associated with $N = 150$ are not consistent with the yield patterns for other levels of fertilization. There is no reason to expect average yields to decline when $N = 150$ and W is increased from 27.2 to 33.1 acre-inches, particularly since the average yield for I_5 is very high.

$$\text{yield} = -2402.5821 + 335.8473 W_1 + 9.8284^{**} N - 26.9328 W_1^{1.5}$$
$$(3037.50) \quad (249.53) \quad (4.478) \quad (27.16)$$

$$- .6480^{***} N^{1.5} + .1427^{*} W_1 N \qquad\qquad (7.7)$$
$$(.203) \quad (.075)$$

$$R^2 = .844 \qquad F = 41.24^{***} \qquad LOFF = 7.89^{**}$$

Consequently, the four observations generated by $W = 33.1$ and $N = 150$ were omitted, and the polynomial forms rerun. The 1.5 model based on the remaining 40 yield observations is in (7.8). The new $LOFF$ is barely significant at the 5 percent level; the tabled F is 2.45. Compared to (7.7), estimated coefficients for the W_1 variables are highly significant. Other test statistics have also improved. Using (7.8) as a predicting equation, derived yields for alternative water and nitrogen levels are given in Table 7.4.

$$\text{yield} = -5949.7309^{**} + 638.2958^{***}W_1 + 15.7148^{***}N - 60.3808^{***}W_1^{1.5}$$
$$\qquad (2405.90) \qquad (198.19) \qquad (3.581) \qquad (21.60)$$

$$\qquad - .9888^{***}N^{1.5} + .1446^{**}W_1 N \qquad\qquad (7.8)$$
$$\qquad\quad (.167) \qquad\quad (.057)$$

$$R^2 = .915 \qquad F = 73.00^{***} \qquad LOFF = 2.468^{**}$$

YUMA MESA, ARIZONA, 1971–72

Inia 66 was sown at the rate of 100 pounds per acre on Superstition Fine Sand at the Yuma Mesa site. Two hundred pounds of $P_2 O_5$ per acre were applied during seedbed preparation. Four inches of water were applied for germination and initial plant growth. Irrigation treatments $I_1, \ldots, I_5$ were applied when the estimated soil moisture level fell to 4.5, 5.16, 5.82, 6.48, and 7.14 percent, respectively. No significant amount of rainfall was recorded. Soil at the Yuma Mesa site has an estimated available water-holding capacity of only 3.1 acre-inches in the upper 4 feet. The soil texture comprises an average 70.9 percent sand, 21.3 percent silt, and 7.8 percent clay.

Because soil nutrients are easily leached at this site, the nitrogen applications were made with water treatments. Nitrogen applications were begun with the first irrigations and continued until plants were entering the boot stage. Treatment combinations and corresponding yields are summarized in Table 7.5.

The quadratic function fitted to the 66 observations is given in (7.9). Few coefficients are statistically significant at conventional levels. The $LOFF$ statistic is just significant at the 5 percent level.

$$\text{yield} = 611.5972 - 57.0650W_1 + 9.8719N + 3.1469W_1^2 - .0569^{***}N^2$$
$$\qquad (1312.39) \quad (163.74) \quad (6.453) \quad (4.747) \quad (.012)$$

$$\qquad + .6092^{**}W_1 N \qquad\qquad (7.9)$$
$$\qquad\quad (.271)$$

$$R^2 = .841 \qquad F = 63.50^{***} \qquad LOFF = 2.223^{**}$$

The negative and positive signs for the linear and quadratic water variables, respectively, cause MP_W to have a positive slope.

When the $(Y - \hat{Y})$ residuals are plotted against W_1 and N, the residuals associated with the $N = 125$ observations are not symmetrically distributed about a zero line. When these six observations are omitted, the quadratic function takes the form

$$\text{yield} = -27.5214 + 29.6823W_1 + 8.9308N + .4544W_1^2 - .0629^{***}N^2$$
$$(1221.37) \quad (151.98) \quad (5.863) \quad (4.397) \quad (.011)$$
$$+ .7462^{***}W_1N \tag{7.10}$$
$$(.249)$$
$$R^2 = .876 \quad F = 76.27^{***} \quad LOFF = .888$$

The *LOFF* statistic is improved. Coefficients for the W_1 variable are both positive but not statistically significant. When a level of fertilizer other than $N = 0$ is selected and average yields are plotted against the water treatment levels, the average yield response curves to water are either nearly linear or curvilinear but convex to the origin. Yield response curves of the latter form are generated when the level of nitrogen is high relative to water applied and production is occurring in stage one of the production function, the stage of increasing returns to water. Consequently, the quadratic model in (7.10) was modified by omitting the W_1^2 variable. The reestimated model is in (7.11). The coefficient for W_1 is statistically significant at about the 11 percent level. Because of the linear relationship for water implicit in (7.11), this equation should not be used for projecting yields beyond the range of observed data. Predicted yields for alternative water and nitrogen combinations are given in Table 7.6.

$$\text{yield} = -143.8832 + 45.1071W_1 + 8.4521^{**}N - .0626^{***}N^2 + .7678^{***}W_1N$$
$$(468.52) \quad (28.26) \quad (3.561) \quad (.011) \quad (.134)$$
$$R^2 = .876 \quad F = 97.09^{***} \quad LOFF = .742 \tag{7.11}$$

YUMA MESA, ARIZONA, 1970-71

Seed variety and seeding rate were identical to the 1971-72 experiment. Prior to planting, 200 pounds of P_2O_5 per acre and 25 pounds of nitrogen per acre were applied to the experimental plots. Ten acre-inches of water were applied for germination and initial plant growth. Irrigation treatments $I_1, \ldots, I_5$ were made when estimated soil moisture dropped to 4.5, 5.1, 5.6, 6.2, and 6.7 percent, respectively. No appreciable amount of rainfall occurred during the growing season. Five levels of nitrogen ranging from 25 to 325 pounds of N per acre were applied. Treatment levels and corresponding yields are summarized in Table 7.7.

Based on an analysis of variance, there was no block effect for this experiment. Among the quadratic, square root, and 1.5 polynomials, the square root function in (7.12) had the lowest *LOFF* statistic. The statistic, however, is significant at the 5 percent level. Consequently, the square root function does not adequately represent yield-water-nitrogen relationships for this experiment. The yields in Table 7.7 were plotted against water treatment levels for each level of N. Among the five yield "response to water curves," only the curve for $N = 25$ was not upward sloping and thus convex to the origin. As with the 1971-72 experiment at Yuma Mesa, the level of nitrogen was too high relative to the quantity of water applied.

$$\text{yield} = 2537.1288^{***} + 83.1833^{*}W_1 + 1.8624N - 696.8977^{***}W_i^{.5}$$
$$(343.07) \qquad (42.56) \qquad (5.685) \qquad (174.43)$$

$$- 351.2444^{***}N^{.5} + 70.8486^{**}W_i^{.5}N^{.5} \qquad\qquad (7.12)$$
$$(98.63) \qquad\qquad (26.37)$$

$$R^2 = .943 \qquad F = 126.68^{***} \qquad LOFF = 4.280^{**}$$

When the $(Y - \hat{Y})$ residuals are plotted against W_1 and N, the observations associated with W_1 = 25.6 and 35.2 are not well distributed about a zero line. When these observations are omitted, the signs and magnitudes of the coefficients for the square root form were such that predicted yields decreased, reached a minimum, and then began increasing. This outcome was also generated with the cubic form.

The final approach considered was deletion of the observations associated with $N = 25$. When N is fixed at 25 pounds per acre, average yield response to higher levels of water applied is negative but also minimal. Several polynomial forms were fitted to the remaining 32 observations. A modified quadratic in (7.13) with the linear N variable omitted seemed most appropriate for the data. Predicted yields derived from (7.13) are listed in Table 7.8.

$$\text{yield} = -3681.4691^{**} + 237.1565^{**}W_1 - 2.6360W_1^2 - .0038N^2 + .1820^{*}W_1N$$
$$(1678.52) \qquad (105.33) \qquad (1.640) \qquad (.005) \qquad (.098)$$

$$R^2 = .907 \qquad F = 65.51^{***} \qquad LOFF = 3.262^{**} \qquad\qquad (7.13)$$

MESA, ARIZONA, 1971–72

A spring wheat variety Siete Cerros was planted November 19, 1971, at a rate of 55 pounds per acre on Laveen Clay Loam at Mesa, Arizona. A preplant irrigation of 8 acre-inches raised the soil moisture level to or near field capacity. The soil at this site has an estimated available water-holding capacity of 7.3 acre-inches in the upper 4 feet.

Irrigation scheduling was based on percentage soil moisture and percentage of available soil moisture used in the upper 3 feet of soil. The five irrigation treatments were based on the following soil moisture relationships:

	Soil moisture	Available moisture used	Soil moisture tension
	(percent)	(percent)	(atmospheres)
I_1	8.5	95	12.0
I_2	9.8	82	5.0
I_3	11.0	70	2.0
I_4	12.3	60	.9
I_5	13.6	40	.7

Record high temperatures for three weeks in March and excessive vegetative growth in plots receiving relatively high applications of nitrogen increased water use by plants above normal levels. Rainfall was not an important factor in this experiment.

The nitrogen treatments were made in three equal applications. The first was made at planting, the second at the "jointing" stage, and the third near the "flowering" stage. Sixty pounds of nitrogen per acre were applied to crop residue in June 1971. In addition, 200 pounds of treble super phosphate per acre were applied in October 1971.

Treatment levels and corresponding yields are given in Table 7.9. No block effect between the two blocks was observed for this experiment. The quadratic function fitted to these data is in (7.14). Both coefficients for the nitrogen variables have negative coefficients.

$$
\text{yield} = -6604.8503^{***} + 941.0952^{***}W_1 - 6.0162N_1 - 16.8489^{***}W_1^2
$$
$$
(1832.67) \qquad (140.41) \qquad (4.286) \qquad (2.675)
$$
$$
- .0085N_1^2 + .3150^{**}W_1N_1 \tag{7.14}
$$
$$
(.0066) \qquad (.141)
$$
$$
R^2 = .588 \qquad F = 10.82^{***} \qquad LOFF = 4.044^{**}
$$

Neither is statistically significant at the 10 percent or lower level. Referring to Table 7.9, the yield response to nitrogen is not strong. For I_4 and I_5, the response is negative. Average yields associated with I_5 exhibit an unusual trend in declining and then increasing when N is varied from 0 to 160 to 320 pounds per acre. Yields-associated I_5 and $N = 160$ are not consistent with the other yield-water-nitrogen relationships. A plot of the $(Y - \hat{Y})$ residuals against W_1 indicates that the residuals are reasonably well distributed. A modified quadratic form omitting the N_1^2 variable was also estimated. The LOFF statistic in (7.15) is still significant at the 5 percent level. All estimated coefficients are statistically significant at the 10 percent or lower level. When the water variables are defined in terms of water used, W_2, estimated parameters for a square root function are given in (7.16).

$$
\text{yield} = -6119.3285^{***} + 910.3067^{***}W_1 - 7.8621^{***}N_1 - 16.0648^{***}W_1^2
$$
$$
(1808.16) \qquad (139.50) \qquad (4.071) \qquad (2.626)
$$
$$
+ .2512^{*}W_1N \tag{7.15}
$$
$$
(.133)
$$
$$
R^2 = 570 \qquad F = 12.90^{***} \qquad LOFF = 3.860^{**}
$$
$$
\text{yield} = -53843.0480^{***} - 2463.5845^{***}W_2 - 10.3014^{*}N_1 + 24217.6258^{***}W_2^{.5}
$$
$$
(9478.86) \qquad (435.29) \qquad (5.388) \qquad (4035.26)
$$
$$
- 557.6692^{**}N_1^{.5} + 153.9061^{***}W_2^{.5}N_1^{.5} \tag{7.16}
$$
$$
(277.83) \qquad (56.75)
$$
$$
R^2 = .547 \qquad F = 9.175^{***} \qquad LOFF = 4.877^{**}
$$

Predicted yields estimated from (7.15) are listed in Table 7.10.

MESA, ARIZONA, 1970-71

Siete Cerros was planted at a rate of 50 pounds per acre. A preplant irrigation of 8 acre-inches raised the soil moisture level to, or near, field capacity. Irrigation scheduling was essentially identical to that for the 1971-72 experiment. Rainfall during the growing season was negligible.

The fertilization levels were slightly changed from the 1971-72 experiment. Estimates of preplant or residual nitrogen are not available. As with the 1971-72 experiment, the nitrogen treatments were made in three equal applications at "planting," "jointing," and "flowering." Two hundred fifty pounds of treble super phosphate per acre were applied prior to planting

Treatment levels and corresponding yields are listed in Table 7.11. At the $N = 0$, 150, or 300 levels, average yields decrease when water is increased from I_3 to I_5. No block effect was observed between the two blocks for this experiment. The 1.5 polynomial in (7.17) was chosen to represent this experiment. All estimated coefficients are highly significant. The *LOFF* statistic, however, is also significant at the 5 percent level. The distributions of the $(Y - \hat{Y})$ residuals did not provide a basis for deleting observations to improve the *LOFF*.

$$\text{yield} = -12523.3684^{***} + 1772.0763^{***}W_1 + 52.7706^{***}N - 229.8171^{***}W_1^{1.5}$$
$$(3667.91) \qquad (460.45) \qquad (4.818) \qquad (62.79)$$

$$- 3.3376^{***}N^{1.5} + .6017^{**}W_1N \qquad\qquad (7.17)$$
$$(.322) \qquad\quad (.250)$$

$$R^2 = .938 \qquad F = 114.55^{***} \qquad LOFF = 4.724^{**}$$

Measurements of available soil moisture at harvest were also made. When W_2 replaces W_1 in a 1.5 polynomial, (7.18) is generated. In comparison with (7.17), the *LOFF* is considerably higher. The other properties are relatively similar. Predicted yields derived from (7.17) are summarized in Table 7.12.

$$\text{yield} = -8257.8517^{**} + 1284.5554^{***}W_2 + 50.9058^{***}N - 165.8639^{***}W_2^{1.5}$$
$$(3224.76) \qquad (420.67) \qquad (4.983) \qquad (58.63)$$

$$- 3.0772^{***}N^{1.5} + .4877^{*}W_2N \qquad\qquad (7.18)$$
$$(.363) \qquad\quad (.285)$$

$$R^2 = .930 \qquad F = 100.28^{***} \qquad LOFF = 6.278^{**}$$

SAFFORD, ARIZONA, 1970-71

Wheat variety Nugaines was broadcast at a rate of 100 pounds per acre on Pima Clay Loam Variant. Irrigation treatments $I_1, \ldots, I_5$ were made when the level of available soil moisture was depleted to an average of 85, 75, 65, 55, and 45 percent, respectively. The soil at this site has an estimated available water-holding capacity of 7.5 acre-inches in the top 4 feet. No significant amount of rainfall was recorded during the growing season. Five levels of nitrogen were incorporated in the experiment. N_1 and N_2 were applied prior to planting. Nitrogen applications were split for N_3-N_5. One-half was applied prior to planting and the remaining half when plants were at the "early stooling" stage. Measurements of preplant N are not available.

Treatment levels and corresponding yields are given in Table 7.13. Except for $W = 27.9$, average yields in Table 7.13 are declining for each level of water applied when the rate of nitrogen is increased. When $W = 27.9$, average yields substantially decrease and then slightly increase when N is raised from 0 to 150 to 300 pounds per acre. The curves representing yield response to water are increasing at all nitrogen rates.

Based on an analysis of variance, Block 2 was significantly different from Block 1 at the 5 percent level. This block effect may be due to soil heterogeneity and (or) differences in residual soil fertility generated by previous cropping patterns. The quadratic form including variable X_1 to represent the block effect is in (7.19). Only the coefficients for X_1 and W are statistically significant at the 5 percent or lower level. As would be expected from the relationships between average yields and levels of fertilization in Table 7.13, the coefficient for N is negative. The $(Y - \hat{Y})$ residuals were plotted against W and N and the residuals were reasonably distributed about a zero line for both inputs. Predicted yields derived from (7.19) are listed in Table 7.14.

$$\text{yield} = -3668.1007 + 525.3182^{**}X_1 + 298.6542^{**}W - 2.6733N$$
$$\quad\quad (2651.76) \quad\quad (199.92) \quad\quad (130.83) \quad\quad (3.892)$$

$$\quad -2.3137W^2 + .0067N^2 - .0675WN$$
$$\quad\quad (1.440) \quad\quad (.010) \quad\quad (.060)$$

$$\tag{7.19}$$

$$R^2 = .800 \quad\quad F = 24.66^{***} \quad\quad LOFF: \text{Block 1} = 1.162$$
$$\quad\quad\quad\quad\quad\quad\quad\quad\quad\quad\quad\quad\quad\quad\quad\quad \text{Block 2} = 0.313$$

WALSH, COLORADO, 1970-71

Wheat variety Scout 66 was planted at a rate of 60 pounds per acre on Baca Clay Loam at Walsh, Colorado. Land at the experimental site was fallowed during the previous year. This soil has an estimated available water-holding capacity of 6.3 inches in the upper 4 feet.

Combinations of three irrigation and five nitrogen treatments were randomly assigned to two blocks each containing 24 experimental plots. Irrigation scheduling was based on electrical resistance block readings within the top foot of soil. Irrigation treatments for I_1, I_2, and I_3 were made when estimated soil water tension in the upper foot of soil reached 1, 3, and 6 bars, respectively. Rainfall exceeding 0.25 inch at any occurrence totaled 7.5 inches during the growing season.

Treatment levels and corresponding yields are given in Table 7.15. A statistically significant block effect between Blocks 1 and 2 was observed but only at the 10 percent level. Consequently, a variable for block effect is not included in the quadratic formulation in (7.20). Only the estimated coefficient for the intercept is statistically significant at conventional levels. Both linear and quadratic terms for N have negative coefficients. Referring to

Table 7.15, average yield response to increased nitrogen does not exhibit any stable relationship or trend.

$$\text{yield} = 3802.4254^{***} + 37.0514W_1 - 1.9692N - .3525W_1^2$$
$$\quad\quad (922.23) \quad\quad (126.42) \quad\quad (3.320) \quad (4.161)$$

$$\quad\quad - .0067N^2 + .2236W_1N \quad\quad\quad\quad\quad (7.20)$$
$$\quad\quad (.011) \quad\quad (.160)$$

$$R^2 = .310 \quad\quad F = 3.77^{***} \quad\quad LOFF = 0.550$$

Measurements of soil moisture at harvest were also made. The quadratic formulation incorporating W_2 is given in (7.21). No improvement over (7.20) is evident. None of the estimated coefficients are statistically significant at the 10 percent or lower levels.

$$\text{yield} = 3156.3033 + 80.5615W_2 - 3.5828N - 1.2114W_2^2 - .0067N^2$$
$$\quad\quad (2256.84) \quad (225.55) \quad\quad (4.369) \quad\quad (5.529) \quad\quad (.011)$$

$$\quad\quad + .2463W_2N \quad\quad\quad\quad\quad\quad\quad (7.21)$$
$$\quad\quad (.184)$$

$$R^2 = .307 \quad\quad F = 3.73^{***} \quad\quad LOFF = 0.566$$

Predicted yields derived from (7.20) are given in Table 7.16. The signs and coefficients of the variables in (7.20) are such that a maximum yield does not exist. Table 7.16 is valid only for the range of water and fertilizer levels in the experiment. Recall W ranged from 10 to 20 acre-inches and N from 0 to 200 pounds per acre.

ANALYSIS OF COTTON EXPERIMENTS

Cotton is an important irrigated crop where growing conditions are favorable, particularly in areas of the Southeast and Southwest. Field experiments for estimating yield-water-nitrogen relationships were conducted at six sites, two in California and four in Arizona. Even though the geographical representation of the experiments is limited, soil characteristics at the experimental sites are of sufficient variation for examining the impact of these site variables on yield. Three years of observations are available for each of the two California sites. Single experiments were conducted at each site in Arizona. The limitations on interpreting results from single-year experiments must be kept in mind. A randomized incomplete block design with factorials was used for all experiments except those completed in California in 1967 and 1968. For these, a central composite rotatable design was used.

Since cotton bolls do not mature evenly, two or more harvests were made on each plot at each site. Mechanical pickers will not harvest immature bolls; at some sites, bolls are handpicked. Plot yields represent the sum of these successive harvests. Yields for the California sites are in pounds of lint cotton per acre. For the Arizona sites, yields are expressed in pounds of seed cotton per acre along with corresponding lint equivalents. Production functions and subsequent analyses are always based on yields of lint cotton. These yields are directly related to the number of plant flowers, the proportion of flowers retained for producing bolls, size of bolls, and lint equivalent of seed cotton produced.

The format for discussing the experiments and presenting analyses of the data is similar to that for the crops covered earlier. To keep the discussion to manageable length, the focus is on yield. Additional data such as boll size, seeds per boll, fiber length and quality, and nitrate nitrogen levels in cotton petioles are available for most experiments. These data, however, are not included in this study. The 1969 experiment at West Side, California, will be analyzed in detail to illustrate the various types of information derivable from these field experiments. As before, the scope of discussion for individual experiments is conditioned by the amount of information available.

SHAFTER, CALIFORNIA, 1969

Acala SJ-1 was planted at a rate of 22 pounds per acre in 40-inch rows on Hesperia Sandy Loam at the United States Cotton Research Station. Following emergence, plants were thinned to a population approximating 19,000 per acre. A preplant irrigation raised the available soil moisture level to an equivalent of 5 inches in the top 5 feet of soil. The soil at the Shafter site has an estimated available water-holding capacity of 7.8 inches in the upper 4 feet of soil. Irrigation scheduling and quantities applied were determined by considering levels of available soil moisture and morphologic stages of plant development. Fertilizer treatments were applied as a side-dressing in late May.

Thirteen water and fertilizer treatment combinations, some duplicated, were randomly applied to a total of 22 plots in each block. Treatment levels and corresponding plot yields adjusted to per acre levels are summarized in Table 8.1.

Choosing any level of nitrogen, average yields increase with higher levels of water applications. The yield response to increased levels of fertilization at any level of water is not substantial. This is likely partly due to the high nitrate content (25 ppm) of the water applied. This yield-nitrogen relationship will be evident in the production function fitted to these data.

Based on an analysis of variance, no block effect between Blocks 1 and 2 was observed. Among the commonly used polynomials, the quadratic function in (8.1) appeared slightly superior for quantifying yield-water-nitrogen relationships for this experiment. The coefficient for W_1 where W_1 = water applied plus 5 inches preplant available soil moisture plus 1 inch rainfall is statistically significant at the 5 percent level. Although of the correct signs, none of the coefficients involving N are significant at conventional levels. The estimated coefficient for the interaction variable is positive. Grimes and Hagan postulate that "a negative interaction is more nearly representative of a normal situation with cotton since excessive quantities of these production factors may result in an unfavorable vegetative-fruiting balance."[1] Finally, the $LOFF$ statistic is not significant at the 5 percent level.

Using (8.1) as a predicting equation, the derived yields corresponding to various levels of W_1 and N are summarized in Table 8.2. The yield response to nitrogen can be examined by choosing a level of water and then reading horizontally across that row. The response to nitrogen applications is low. Recall that the water applied had a relatively high nitrate content. The average level of W_1 in this experiment was close to 30 inches. Setting $W_1 = 30$ in Table 8.2, predicted yields increase from 779 to only 845 pounds per acre when the nitrogen treatment is increased from zero to 240 pounds per acre. This is the range of fertilizer applications in the experiment. The average product of one pound of nitrogen through this range is (845-779)/240 = 0.275 pound of lint cotton per pound of nitrogen. The estimated average product of an acre-inch of water is considerably higher.

$$\text{yield} = 233.7181 + 23.6508^{**}W_1 + .4389N - .1820W_1^2$$
$$\quad\quad\ (164.97)\quad\ (11.09)\quad\quad (.844)\quad\ (.174)$$

$$\quad\quad - .0033N^2 + .0209W_1N \quad\quad\quad\quad\quad\quad (8.1)$$
$$\quad\quad\ \ (.003)\quad\quad (.016)$$

$$R^2 = .863 \quad\quad F = 20.11^{***} \quad\quad LOFF = 2.807$$

SHAFTER, CALIFORNIA, 1968

Acala SJ-1 was again planted at a seeding rate of 22 pounds per acre. The only managerial dissimilarities from the 1969 experiment were that the experimental design was a central composite rotatable design instead of an incomplete block and that different levels of water and nitrogen were applied.

Individual and average plot yields of lint cotton adjusted to per acre levels are summarized according to treatment combinations in Table 8.3. Various combinations of five levels of water and nitrogen were used. The yield response to nitrogen when water is fixed at any level is stronger than that observed for the 1969 experiment. When $N = 125$, however, average yield rises as water is increased from 15.75 to 30.75 acre-inches but then starts decreasing at some point beyond $W_1 = 30.75$.

As with the 1969 experiment, no block effect was observed. The implication is that both blocks of experimental plots had essentially homogenous soil characteristics, and they were subjected to nearly identical management practices. A quadratic formulation in (8.2) was fitted to the 26 observations pooled from the two blocks of plots. For this set of data, the negative coefficient for $W_1 N$ indicates that W_1 and N are substitutes. That is, the marginal product of W_1 declines as N is increased and vice versa. The same conclusion applies to the marginal product for N. Based on the $LOFF$ statistic, the quadratic model seems to adequately fit the data being analyzed.

$$\text{yield} = -1103.6227^{***} + 118.3453^{***}W_1 + 2.8459^{**}N - 1.6287^{***}W_1^2$$
$$\quad\;\;(227.15)\qquad\;\;(12.89)\qquad\;(1.220)\qquad\;(.197)$$

$$- .0040N^2 - .0461W_1N \tag{8.2}$$
$$\;\;(.003)\qquad(.031)$$

$$R^2 = .850 \qquad F = 22.75^{***} \qquad LOFF = .441$$

Predicted yields derived from (8.2) are listed in Table 8.4. Note the negative yield associated with $W_1 = 10$ and $N = 0$. The model implicit in 8.2 was fitted to water levels ranging from 15.75 to 45.75 acre-inches and nitrogen from 0 to 250 pounds per acre. Projecting predicted yields beyond these ranges can generate unreasonable results. The negative yield in Table 8.4 is an example of an inappropriate projection.

Yield response curves to water and (or) nitrogen can be readily derived from Table 8.4. Predicted yields begin decreasing when W_1 is in the range of 35 to 40 acre-inches. In this experiment, yield is maximized when W_1 and N are in the vicinity of 35 acre-inches and 150 pounds per acre, respectively.

SHAFTER, CALIFORNIA, 1967

The 1967 experiment was essentially identical to the 1968 trial. Yields and treatment levels are summarized in Table 8.5.

No block effect was evident for this experiment. Based on test statistics, including the $LOFF$, the square root function in (8.3) is suitable for this set of data. Coefficients for the intercept and water variables are statistically significant at the 10 percent or lower level and signs on other variables are consistent with expectations.

$$\text{yield} = -4547.1232^{***} - 154.9430^{***}W_1 - .9533N + 1838.9345^{***}W_1^{.5}$$
$$\quad\quad (712.92) \quad\quad (20.98) \quad\quad (.720) \quad\quad (235.24)$$

$$+ 42.4257N^{.5} - 4.4352W_1^{.5}N^{.5} \quad\quad\quad\quad (8.3)$$
$$\quad (37.16) \quad\quad (6.338)$$

$$R^2 = .853 \quad F = 23.22^{***} \quad LOFF = 1.511$$

As before, the negative coefficient for the interaction term indicates that W_1 and N are substitutes in cotton production. Predicted yields derived from (8.3) are in Table 8.6. Predicted yield reaches a maximum when W_1 and N are in the vicinity of 35 acre-inches and 50 pounds per acre, respectively. The low level of N, which maximizes yield, may result from preplant or residual fertilizer not included in the analysis. Also, recall that irrigation water at the Shafter site has a relatively high nitrate content.

SHAFTER, CALIFORNIA, 1967, 1968, AND 1969 COMBINED

The three experiments at the Shafter site are sufficiently similar so that a production function combining the three years of data is of interest. All experiments were conducted at the same site, with essentially the same management input and with the same seed and seeding rates. Recall that the experimental design for the 1967 and 1968 experiments differs from that for 1969. The year-to-year variation in yields is hypothesized to be a consequence of environmental factors, primarily climatic conditions. A review of Tables 8.1, 8.3, and 8.5 shows that the water levels for the three years were in about the same range for the five respective fertilizer levels. The nitrogen levels are also reasonably comparable.

Two sets of variables for incorporating the interyear effect were considered. The first was a weather variable defined as the sum of daily pan evaporation readings from planting up to the date of the first harvest. This procedure is discussed in more detail in Chapter 10 where generalized production functions are developed. In brief, pan evaporation (*PE*) data embody the influences of temperature, humidity, wind, and radiation. As such, these data should represent a composite variable for climatic conditions. Rainfall is considered separately. Letting *PE* represent the weather variable, the following values were estimated:

PE

1967 57.1 inches

1968 65.8 inches

1969 68.3 inches

The second set of variables consists of dummy variables to capture the inter-year variation. The dummy variables would reflect the effect of weather and other environmental conditions as well as variations in management and experiment design. The dummy variables were defined as

$$X_1 \quad X_2$$

	X_1	X_2
1967	1	0
1968	0	1
1969	0	0

When the experimental data are pooled and PE is included, the 1.5 polynomial in (8.4) generates somewhat better results than the quadratic or square root functions. The PE variable is statistically significant at the 5 percent level as are the intercept and water variables.

$$\text{yield} = -1255.\overset{***}{2336} + 6.\overset{**}{707}PE + 145.\overset{***}{5428}W_1 + 1.\overset{*}{8458}N$$
$$\qquad\quad (273.45) \qquad (2.801) \qquad (17.02) \qquad (1.056)$$
$$- 16.\overset{***}{0407}W_1^{1.5} - .100\overset{*}{8}N^{1.5} + .0011W_1N \qquad (8.4)$$
$$\quad (2.043) \qquad\quad (.056) \qquad\quad (.017)$$
$$R^2 = .712 \qquad F = 2\overset{*\,**}{7.65}$$

When the dummy variable approach is considered, the 1.5 polynomial is still appropriate and takes the form represented in (8.5). Given the manner in which the dummy variables were specified, the intercept in (8.5) represents the year effect for 1969. The coefficient for X_1 reflects the difference in "year effect" between 1969 and 1967. Based on the estimated coefficient, this difference was not statistically significant at predetermined levels. Similarly, the coefficient for X_2 estimates the difference in "year effect" between 1969 and 1968. This difference was statistically significant at the 1 percent level. Predicting equations for individual years can be derived by substituting the appropriate values for X_1 and X_2 into (8.5). A predicting equation representing an average over the three years is derived by averaging the intercepts for the three years and then adding the W and N variables of the model. The average intercept in (8.5) is -678.6777; the other coefficients and variables are unchanged. Replacing the intercept, X_1, and X_2 variables in (8.5) by -678.6777, the transformed equation is used to predict the yields appearing in Table 8.7. Yield reaches a maximum in the area of $W = 35$ acre-inches and $N = 200$ pounds per acre.

$$\text{yield} = -703.\overset{***}{8635} - 24.0782X_1 + 99.\overset{***}{6355}X_2 + 131.\overset{***}{1167}W_1$$
$$\qquad\quad (162.82) \qquad (31.79) \qquad (31.79) \qquad (16.05)$$
$$+ 1.1405N - 14.\overset{***}{2896}W_1^{1.5} - .0571N^{1.5} + .0011W_1N$$
$$\quad (.985) \qquad\quad (1.928) \qquad\quad (.053) \qquad\quad (.016) \qquad (8.5)$$

$$R^2 = .763 \qquad F = 30.29^{***}$$

WEST SIDE, CALIFORNIA, 1969

Acala SJ-1 was planted in 40-inch rows at a rate of 30 pounds per acre on Panoche Clay Loam at the West Side Field Station, Fresno County, California. Following emergence, plants were thinned to approximately 17,000 per acre.

A preplant irrigation raised the soil moisture level in the top 6 feet of soil to an equivalent of 10 inches per acre. The available water-holding capacity of the upper 4 feet of soil is estimated at 8.3 inches. Comments on irrigation scheduling and nitrogen applications for the 1969 experiment at Shafter also apply to this experiment.

Two blocks each containing 13 plots were subjected to the water and nitrogen treatments specified in Table 8.8. When nitrogen is applied at a rate of 350 pounds per acre, yields associated with high levels of water application are declining. In this area, the marginal product of nitrogen is negative.

Based on an analysis of variance, no block effect was observed. The square root function in (8.6) was fitted to the 26 yield observations. With the exception of $N^{.5}$, all estimated coefficients in (8.6) are significant at an 0.01 probability level. Even though most properties of (8.6) seem very good, the *LOFF* statistic is significant at the 5 percent level. The tabled value for $F_{(7,13)}$ when $\alpha = 0.05$ is about 3.50. That is, when $\alpha = 0.05$, the square root function is not adequate for quantifying the yield-water-fertilizer relationships represented by the 26 observed yields. The $(Y - \hat{Y})$ residuals were plotted, but no serious deviations from the zero line were observed. Since the estimated *LOFF* is reasonably close to the tabled value, (8.6) is used to represent yield-water-fertilizer relationships for the 1969 experiment at West Side. Using (8.6) as the predicting equation, yields for various combinations of water and fertilizer are given in Table 8.9. Yield reaches a maximum in the vicinity of $W_1 = 38$ acre-inches and $N = 250$ pounds per acre.

$$\text{yield} = -1714.4730^{***} - 78.6963^{***}W_1 - 1.5227^{***}N + 896.2319^{***}W_1^{.5}$$
$$\quad (297.81) \qquad (11.85) \qquad (.430) \qquad (120.96)$$

$$+ 18.1025N^{.5} + 4.6318^{***}W_1^{.5}N^{.5} \qquad\qquad (8.6)$$
$$\quad (11.47) \qquad\quad (1.588)$$

$$R^2 = .938 \qquad F = 60.92^{***} \qquad LOFF = 3.552^{**}$$

Several other physical and economic relationships can be derived from (8.6). A three-dimensional production surface is plotted in Figure 8.1 by substituting values for W_1 and N into (8.6). Yield reaches a maximum at about $W = 35$ acre-inches and then begins to decline. Yield response curves to water for three levels of N are drawn in Figure 8.2. The marginal product of water at any point on one of these curves also represents the value of the

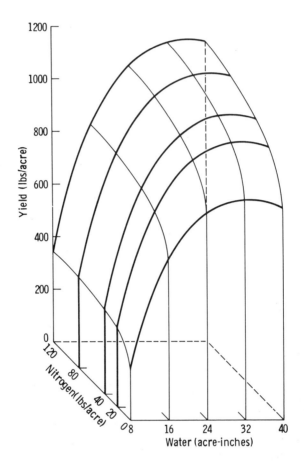

FIG. 8.1. Production surface for lint cotton estimated
from (8.6) according to specified water and
nitrogen levels (West Side, California, 1969).

slope of the curve at that point. When the marginal product of nitrogen is
zero, yield is at a maximum for the specified level of N. When $N = 0$ or 100,
yield is maximized when about 34 inches of water are applied. When N is
increased to 200 pounds per acre, yield is maximized when water applied
equals about 38 inches. This information for $N = 100$ and 200 is also given
in Table 8.9. Applying water beyond the point where its marginal product is
zero is economically irrational. The exception is if additional water is applied
to leach salt accumulations from the root zone.

The difference in yield response in Figure 8.2 when $N = 100$ and $N = 200$
is of interest. When $N = 200$ and $W = 10$, too much fertilizer has been
applied and MP_N is negative. This relationship is also depicted in Figure 8.4
where $MP_N \simeq 0$ when $N = 200$ and $W = 30$. If $W < 30$ and $N = 200$, $MP_N < 0$.

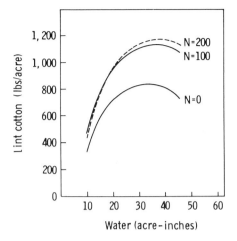

FIG. 8.2. Lint cotton yield response curves to water derived from (8.6) assuming specified levels of nitrogen (West Side, California, 1969).

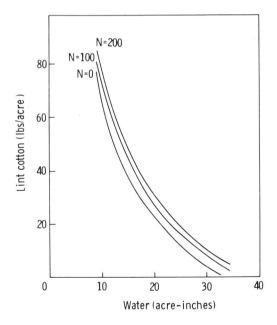

FIG. 8.3. Marginal product curves for water in production of lint cotton at specified levels of nitrogen (West Side, California, 1969).

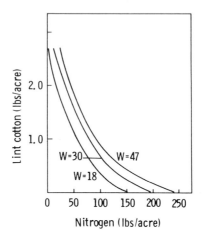

FIG. 8.4. Marginal product curves for nitrogen in production of lint cotton at specified levels of water
(West Side, California, 1969).

The MP_W curves at three levels of fertilization are shown in Figure 8.3.
When $N = 0$, MP_W becomes negative when $W \simeq 32$. At higher levels of fertilization, $MP_W < 0$ when water approaches 38 inches per acre. The positioning
of the curves for $N = 100$ and $N = 200$ indicates that the MP_W is only slightly
increased even though the rate of nitrogen is doubled. This is consistent with
the relationships implicit in Figure 8.2. Marginal product curves for nitrogen
at three levels of water are shown in Figure 8.4. When $N = 100$ and W is
increased from 18 to 30 to 42 acre-inches, MP_N increases from only 0.35 to
0.65 to 0.90 pounds per acre, respectively. The MP_N when N is close to zero
is very high at all levels of W but rapidly diminishes as the rate of nitrogen
application increases.

Marginal product curves for land (MP_L) are plotted in Figure 8.5. The
estimating procedure is identical to that in (6.9) where the marginal product
of land in corn production was estimated. Three levels of water and nitrogen
availability have been assumed for application over successively larger areas
of land. When $W = 375$ acre-inches and $N = 1250$ pounds, MP_L declines at a
diminishing rate and becomes zero when these inputs are spread over about
27 acres of land. When land is fixed at, for example, 20 acres, MP_L increases
from 297 to 627 to 1154 pounds of lint cotton when W and N are at the
levels specified in Figure 8.5.

Isoquants for four levels of yield are plotted in Figure 8.6. Recall that
points on an isoquant represent the MRS between the specified inputs. The
MRS of W for N is extremely high until the level of N is less than 50 pounds
per acre. As a consequence, relatively small increments of water can replace
or be substituted for large quantities of nitrogen without changing per acre
yield. Consider isoquant $Y = 600$ in Figure 8.6. If water is increased only
slightly from about 11.5 .to 11.8 acre-inches, nitrogen can be reduced from

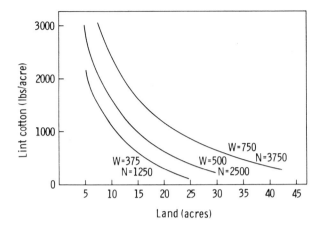

FIG. 8.5. Marginal product curves for land in production
of lint cotton with specified availability of water
and nitrogen (West Side, California, 1969).

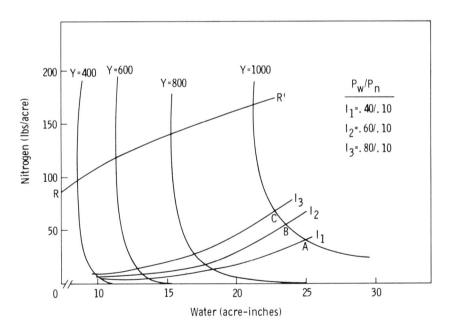

FIG. 8.6. Isoquants and isoclines for specified per acre
yields of lint cotton and price ratios (West Side,
California, 1969).

190 to 50 pounds per acre, and $Y = 600$ is still produced. At lower levels of fertilization, the MRS of water for nitrogen substantially declines and successively larger quantities of water are necessary for replacing an increment, that is, a pound of nitrogen. Beyond the point where $N = 130$, increased amounts of both water and nitrogen are necessary to generate $Y = 600$. At this point, the isoquant tends to become concave to the origin and the marginal rate of substitution is greater than zero. Ridgeline RR' is superimposed to delimit the upper limit on the range where $MRS < 0$ on the family of isoquants. This represents the upper limit on the area of rational production. The nature of the isoquants and the scales of the axes in Figure 8.6 are such that the lower limit beyond which $MRS > 0$ does not appear. Consequently, only one ridgeline is drawn.

Three price relationships designated by I_1, I_2, and I_3 isoclines are reflected in Figure 8.6. The price of N is fixed at \$0.10 per pound while the price of water is varied from \$0.40 to \$0.60 to \$0.80 per acre-inch. Recall that isoclines connect points of equal slopes on successive isoquants. I_1 in Figure 8.6, for example, connects those points on the isoquants where the slope equals $0.40/0.10 = 4.0$. At the points of intersection of I_1 with any isoquant, $MRS = (P_w/P_n)$ and the corresponding levels of W and N represent the least-cost combination of producing the output level under consideration. As P_w increases relative to P_n, water becomes expensive relative to nitrogen. With a higher P_w/P_n ratio, point A, for example, on isoquant $Y = 1000$ is no longer optimal. Points to the left of A represent higher MRSs; that is, the slope of the curve is increasing. When $P_w = 0.60$, the least-cost combination of producing $Y = 1000$ at point B is comprised of about 55 pounds of N per acre and 23.75 acre-inches of water. Since the isoclines are curvilinear, the proportion or mix of W and N changes with movement to higher isoquants even though the P_w/P_n ratio is unchanged.

WEST SIDE, CALIFORNIA, 1968

The 1968 experiment differs from the 1969 trial in two respects. First, a different experimental design and different treatment levels were used. Second, an additional treatment represented by variable plant spacing within rows was used. Treatment levels and corresponding plot yields adjusted to the acre level are summarized in Table 8.10. The impact of plant spacing on yield can be examined by considering those yields in Table 8.10 associated with $W = 20$, $N = 175$, and plant spacing (PS) from 1.6 to 8.8 to 16.0 inches. When $PS = 1.6$, the average yield for two observations is 799 pounds per acre. When plant population is reduced to allow 8.8 inches between plants, average yield increases to 1323 pounds per acre. Finally, the average yield for plots with plants 16 inches apart was 1380 pounds per acre. Substantial yield increases were observed when distance between plants was increased from 1.6 to 8.8 inches but only a modest increase when PS was expanded from 8.8 to 16.0 inches. Plant spacing for each water and nitrogen combination averages 8.8 inches.

No block effect was observed among the 36 observations across two blocks but plant population represented by PS was statistically significant at the 5 percent level. Alternative models incorporating different forms of PS were estimated. Equation (8.7) represents a square root function with a linear PS variable. The $LOFF$ statistic is very high. The model implicit in (8.7) was modified by adding another variable defined as $(PS)^{.5}$. The estimated model is given in (8.8). All estimated coefficients are highly statistically significant and have the correct signs. The $LOFF$ statistic is about half that for (8.7), but still highly significant.

$$yield = -764.8279 + 15.7607^{**}PS - 38.9395^{*}W_1 - 1.5152N$$
$$(594.23) \quad (6.893) \quad (21.32) \quad (.958)$$

$$+ 526.6074^{**}W_i^{.5} + 105.1756^{**}N^{.5} - 14.6615^{*}W_i^{.5}N^{.5} \quad (8.7)$$
$$(203.64) \quad (42.02) \quad (8.480)$$

$$R^2 = .398 \quad F = 3.20^{**} \quad LOFF = 18.886^{***}$$

$$yield = -2525.9478^{***} - 165.5840^{***}PS - 61.2430^{***}W_1 - 2.5072^{***}N$$
$$(541.38) \quad (34.58) \quad (15.90) \quad (.714)$$

$$+ 1005.8189^{***}(PS)^{.5} + 709.5290^{***}W_i^{.5}$$
$$(189.83) \quad (150.46) \quad\quad\quad (8.8)$$

$$+ 129.2883^{***}N^{.5} - 15.2105^{**}W_i^{.5}N^{.5}$$
$$(30.56) \quad (6.099)$$

$$R^2 = .700 \quad F = 9.316^{***} \quad LOFF = 9.276^{***}$$

The residuals were examined to see if (8.8) could be improved. No improvement of the $LOFF$ was evidenced. The high $LOFF$ statistic may be partly due to the fact that with exception of the plots subjected to $W = 20$ and $N = 175$, only two observations per treatment combination are available. Using (8.8) as a predicting equation, derived yields for specified water and nitrogen levels are summarized in Table 8.11. Yield reaches a maximum in the vicinity of $W = 18$ acre-inches and $N = 150$ pounds per acre.

WEST SIDE, CALIFORNIA, 1967

Seed, planting rate, and experimental site are the same as for the 1968 and 1969 experiments at West Side. Following emergence, plants were thinned to approximately 17,000 per acre. About 10 inches of soil moisture were available at planting. No significant amount of rainfall was recorded during the growing season. Individual and average plot yields adjusted to acre levels are summarized in Table 8.12.

No block effect was observed in this experiment. As with the 1968 and 1969 experiments, the square root function is used to quantify the yield-water-nitrogen relationships. Estimated coefficients for the N variables were not statistically significant. This result is also implicit in Table 8.12. Fixing a water level and reading upward, yield response to higher levels of nitrogen is minimal. Predicted yields derived from (8.9) are in Table 8.13.

$$\text{yield} = -2434.4034^{**} - 99.3308^{***}W_1 + .1183N + 1228.9843^{***}W_i^{.5}$$
$$(878.74)(21.66)(1.315)(253.09)$$

$$+ 49.9565N^{.5} - 8.7215W_i^{.5}N^{.5} \tag{8.9}$$
$$(52.92)(7.270)$$

$$R^2 = .615 \qquad F = 6.38^{**} \qquad LOFF = .980$$

WEST SIDE, CALIFORNIA, 1967 AND 1969 COMBINED

The introductory comments for discussion of the combined 1967–1969 experiments at Shafter also apply here. Since the 1968 experiment included a plant spacing variable, it is omitted from the analysis of the combined data. A dummy variable approach and specification of weather variables were used to aggregate the two sets of experimental data. In the model incorporating dummy variables, $X_1 = 1$ and 0 for the 1969 and 1967 experiments, respectively. Using pan evaporation (PE) data to represent weather conditions during the growing season, $PE = 67.4$ and 63.6 inches for 1969 and 1967, respectively.

Using PE as the site-year variable, the square root formulation in (8.10) is derived. When the dummy variable approach is used, (8.11) is generated. Equations (8.10) and (8.11) are essentially identical; the only differences occur in the estimated values for the intercept and the site-year variable. If $PE = 67.4$ is substituted into (8.10), the first two terms sum to -2015.63.

$$\text{yield} = 2123.6439^{***} - 61.4136^{***}PE - 91.8515^{***}W_1 - 1.2545^{***}N$$
$$(626.57)(7.134)(10.99)(.380)$$

$$+ 1028.3796^{***}W_i^{.5} + 20.7464^{*}N^{.5} + 3.1669^{*}W_i^{.5}N^{.5} \tag{8.10}$$
$$(113.50)(11.44)(1.679)$$

$$R^2 = .926 \qquad F = 94.48^{***}$$

$$\text{yield} = -1782.2634^{***} - 233.3718^{***}X_1 - 91.8515^{***}W_1 - 1.2545^{***}N$$
$$(295.92)(27.11)(10.99)(.380)$$

$$+ 1028.3796^{***}W_i^{.5} + 20.7464^{*}N^{.5} + 3.1669^{*}W_i^{.5}N^{.5} \tag{8.11}$$
$$(113.50)(11.44)(1.679)$$

$$R^2 = .926 \qquad F = 94.48^{***}$$

The same value is obtained when $X_1 = 1$ is substituted into (8.11) and the first two terms are combined. If (8.11) is used to predict yields for the 1967 experiment, $X_1 = 0$ and the intercept for the predicting equation is -1782.2634. The other variables and coefficients are unchanged. The intercept for the predicting yields for the 1969 experiment is -2015.6352. The average intercept for the two years of combined data becomes -1898.9493. Predicted yields for alternative water and nitrogen combinations are summarized in Table 8.14.

SAFFORD, ARIZONA, 1971

Deltapine 16 was planted on Pima Clay Loam Variant at the Safford Experiment Station. Dry, windy weather was prevalent during the first month of plant growth. The soil has an estimated available water-holding capacity of 7.5 inches in the top 4 feet.

Measurements of available soil moisture and residual nitrogen prior to planting were not made. Water and nitrogen treatment levels along with corresponding yields are summarized in Table 8.15. Applications of $N = 100$, 150, and 200 pounds per acre were split with one-half applied at planting and the rest during July 9–12. The nitrate content of the irrigation water was about 60 ppm. Significant amounts of rainfall during the growing period totaled 2.45 inches.

Based on an analysis of variance, there was no block effect between the two blocks each containing 22 plots. A quadratic function was fitted to the experimental data. Only coefficients for the nitrogen variables in (8.12) are significant at conventional levels. Recall that the water applied had a high nitrate content. Predicted yields derived from (8.12) are given in Table 8.16.

$$\text{yield} = -170.9291 + 32.4521W_1 + 3.7350\overset{**}{N} - .0222W_1^2$$
$$\qquad\quad (875.27) \qquad (67.10) \qquad (1.436) \qquad (1.258)$$

$$\qquad - .0108\overset{***}{N^2} - .0222W_1 N \qquad\qquad\qquad (8.12)$$
$$\qquad\quad (.004) \qquad\quad (.047)$$

$$R^2 = .705 \qquad F = 18.18\overset{***}{} \qquad LOFF = 2.286\overset{*}{}$$

TEMPE, ARIZONA, 1971

Deltapine 16 was planted in 40-inch rows at the Cotton Research Center, Tempe. Soil at the experimental site is quite variable with pockets of sand, gravel, silt, and clay. The available water-holding capacity in the top 4 feet of soil is estimated as 5.6 inches. A preplant irrigation of approximately 6 inches was applied in early March. On May 3, plants were thinned to allow 8–9 inches between plants. Nitrogen was applied as a side-dressing on May 5.

Treatment levels and corresponding yields are summarized in Table 8.17. The general lack of yield response to nitrogen applications suggests that a substantial amount of residual or preplant nitrogen was available for plant growth.

Based on an analysis of variance, a statistically significant difference between the effects of Blocks 1 and 2 on yield was observed. This block effect may have resulted from variability in soil texture. Let X_1 represent the block effect with values one and zero for yields from Blocks 1 and 2, respectively. The 1.5 polynomial in (8.13) represents yield-water-nitrogen relationships for this experiment. Coefficients for the nitrogen variables are not statistically significant at the 10 percent level. The negative and positive signs for N and $N^{1.5}$, respectively, indicate that the yield response curve to nitrogen is downward sloping, reaches a minimum, and then turns upward. This unreasonable phenomenon can also be observed in Table 8.17 by examining average yields associated with $W = 33.4$ and 37.4 when N is varied from 0 to 150 to 300. The coefficient for X_1 reflects the difference between Blocks 1 and 2.

$$\text{yield} = -22039.4673^{***} - 59.4545^{**}X_1 + 1871.1190^{***}W_1 - 1.9485N$$
$$(5865.05) \quad\quad (25.45) \quad\quad (487.63) \quad\quad (1.760)$$

$$- 204.3537^{***}W_1^{1.5} + .0355N^{1.5} + .0332W_1 N \quad\quad\quad (8.13)$$
$$(54.05) \quad\quad (.037) \quad\quad (.044)$$

$$R^2 = .626 \quad F = 10.32^{***} \quad LOFF = .473$$

Using -22069.1946 as an average intercept coefficient across both blocks, predicted yields derived from (8.13) are given in Table 8.18. Using increments of 5.0, when W is less than 30 or greater than 45 acre-inches, predicted yields are negative.

YUMA MESA, ARIZONA, 1971

Stoneville 7-A was planted in 20-inch rows on Superstition Fine Sand at Yuma Mesa, Arizona. A preplant application of 200 pounds P_2O_5 and 25 pounds of nitrogen per acre was made during seedbed preparation. The N treatment levels were divided among irrigations so that a portion of the total fertilizer treatment was applied with each irrigation. This was done to minimize the amount of leaching in this sandy soil. Eight inches of water were applied prior to planting. No significant amount of rainfall occurred during the growing season. Irrigation treatments were made when percentage soil moisture dropped to 6.74, 6.18, 5.62, 5.06, and 4.50, respectively. The available water-holding capacity in the top 4 feet of soil is estimated at 3.1 inches.

Treatment levels and corresponding yields are given in Table 8.19. Each of the two blocks had 22 plots. This sandy soil requires relatively high applications of both nitrogen and water.

The 1.5 polynomial was selected for analyzing this experiment. No block effect between the two blocks was observed. Estimated coefficients for the nitrogen variables are not significant at conventional levels but do have the appropriate signs. Predicted yields derived from (8.14) are listed in Table 8.20.

$$\text{yield} = -1416.2202^{*} + 121.5331^{***}W_1 + .2471N - 10.9715^{***}W_1^{1.5}$$
$$(710.89) \quad\quad (33.44) \quad\quad (1.430) \quad\quad (2.751)$$

$$- .0916^{*}N^{1.5} + .0424^{***}W_1 N \quad\quad\quad (8.14)$$
$$(.052) \quad\quad (.008)$$

$$R^2 = .630 \quad F = 12.96^{***} \quad LOFF = 1.923$$

YUMA VALLEY, ARIZONA, 1971

Stoneville 7-A was planted in 40-inch rows on Glendale Silty Clay Loam at Yuma Valley, Arizona. Prior to planting, 9.6 acre-inches of water and 100 pounds of P_2O_5 per acre were applied. No significant amount of rainfall occurred during the growing season. Irrigation treatments $I_1, \ldots, I_5$ were made when the percent soil moisture dropped to 16.4, 18.0, 19.6, 21.2, and 22.8, respectively. The corresponding levels of percent of available water used are 85, 75, 65, 55, and 45. All the nitrogen for $N = 75$ and 150 was applied with the first irrigation; the $N = 225$ and 300 treatments were split with the two applications 4–6 weeks apart.

Two blocks each containing 22 plots were incorporated into an incomplete block experimental design. Treatment levels and corresponding yields and lint percentages are summarized in Table 8.21. When the rate of fertilization is greater than 150 pounds per acre, average yields decrease at all water levels.

No block effect was observed with this experiment. The square root function was selected for quantifying the yield-water-nitrogen relationships. Even though the $LOFF$ statistic indicates the model is appropriate or plausible for data, the R^2 is also quite low. Predicted yields derived from (8.15) are given in Table 8.22.

$$\text{yield} = -5467.8895 - 113.8108W_1 - 1.8941^{***}N + 1775.1452W_1^{.5}$$
$$(27072.62) \quad (498.99) \quad (.667) \quad (7353.18)$$

$$+ 28.5394N^{.5} + 1.3376W_1^{.5}N^{.5} \tag{8.15}$$
$$(100.47) \quad (13.55)$$

$$R^2 = .300 \quad F = 3.26^{**} \quad LOFF = .966$$

SINGLE PRODUCT AND CANONICAL JOINT PRODUCT FUNCTIONS FOR SUGAR BEETS

Sugar beet experiments were conducted in Arizona, Colorado, and Texas. Root yields and corresponding sucrose percentages are available for all trials and top yields for most. Early and late harvests were made at the Mesa, Arizona, site. The composition of the plant changed during this period. Root yields and sucrose percentages were relatively higher in the late harvest, but top yields were lower. The later harvest also required larger applications of water. Consequently, a substitution existed among water, yields, and sucrose percentages. Production functions are estimated for the root and yield components. The root yields have been adjusted to reflect a standard sucrose percentage of 15 percent.

MESA, ARIZONA, 1971-72

Variety S301-H8, a monogerm, nonbolting hybrid with curly top resistance, was planted September 23, 1971, in 40-inch, double-row beds at the Mesa Branch Experiment Station, Mesa, Arizona. Prior to planting, 200 pounds of 0-45-0 per acre were uniformly applied; 8 acre-inches of water were applied for germination. The Laveen Clay Loam soil at this site has an estimated available water-holding capacity of 7.3 acre-inches in the top 4 feet. Plant seedlings were hand-thinned to 10-inch intervals.

Five irrigation treatments were designed to maintain soil moisture tension below a specified level and to reflect the following relationships:

	Soil Moisture Tension (atmospheres)	Available Moisture Depleted (percent)
I_1	0.4	25
I_2	0.6	35
I_3	1.0	55
I_4	3.0	65
I_5	10.0	90

The number of irrigations varied directly with the level of fertilization. Plots subjected to heavy fertilization also produced heavy top yields. Plants in these plots used available water more rapidly than those receiving lighter applications of fertilizer. Rainfall exceeding 0.25 inch at any occurrence totaled about 1 inch during the growing season.

The level of preplant soil nitrogen in the upper 3 feet of soil was estimated at about 75 pounds per acre. Five levels of fertilization ranging from 0 to 360 pounds of nitrogen per acre were incorporated in the experimental design. Plants in those plots receiving no fertilizer exhibited foliar symptoms of nitrogen deficiency by late November 1971.

Plots were harvested on May 4 and July 7, 1972. These dates correspond to the early and late season commercial harvest periods in central Arizona. Leaves and crown tissue removed in the topping operation were weighed and recorded as tops. A sample of 12 to 20 roots per plot was analyzed for sucrose content at the Spreckels Sugar Company laboratories.

May harvest

Yield-water-nitrogen data and sucrose percentages for the root yields are in Table 9.1. Average yields for roots adjusted to 15 percent sucrose and for tops are also listed. The average yield response to water for roots is weak at all levels of nitrogen while the average yield response for tops to water is more pronounced. The average yield response to nitrogen is much stronger, especially for top yields.

Root yield. Based on an analysis of variance, a block effect existed for these data. Among the quadratic, square root, and 1.5 polynomial forms, (9.1) provides the best fit for quantifying yield-water-nitrogen relationships for roots at 15 percent sucrose. This is consistent with the earlier discussion of average yield responses implicit in Table 9.1. Since the estimated coefficient for the interaction term is negative, water and nitrogen are substitutes in the production of roots for this experiment. The *LOFF* statistics are not significant at the 5 percent level. Consequently, (9.1) is considered adequate or appropriate for quantifying the relationships between root yields and treatments applied. Predicted yields derived from (9.1) are listed in Table 9.2.

$$\begin{aligned} \text{yield} &= -52.4647 + .8516\overset{*}{X}_1 - 1.2792W_1 - .1221\overset{***}{N}_1 \\ \text{(roots)} \quad &(37.10) \quad (.464) \quad (1.643) \quad (.015) \\ &+ 15.4826W_i^5 + 4.0324\overset{***}{N}_i^5 - .0555W_i^5 N_i^5 \\ &(15.74) \quad (1.023) \quad (.178) \\ R^2 &= .759 \quad F = 19.42\overset{***}{} \quad LOFF: \text{Block } 1 = 3.075 \\ & \qquad\qquad\qquad\qquad\qquad\quad \text{Block } 2 = 2.147 \end{aligned} \tag{9.1}$$

Top yield. In analyzing the top yield data, the 1.5 polynomial in (9.2) was derived. There was no block effect among the top yield data. Estimated coefficients for the water variables are highly significant. The strong average yield response to nitrogen implicit in Table 9.1 is apparently captured by the interaction term. Whenever water and nitrogen are both increased, average yields for tops in Table 9.1 always increase. Predicted yields derived from (9.2) are in Table 9.3.

$$\text{yield} = -85.3416^{***} + 10.1131^{***}W_1 + .0188N_1 - 1.2767^{***}W_1^{1.5}$$
$$\text{(tops)} \quad (26.71) \qquad (3.178) \qquad (.045) \qquad (.416)$$

$$- .0026^{*}N_1^{1.5} + .0041^{***}W_1N_1 \qquad\qquad\qquad (9.2)$$
$$\quad (.001) \qquad\quad (.001)$$

$$R^2 = .949 \qquad F = 142.82^{***} \qquad LOFF = 1.964$$

July harvest

Treatment levels, root and top yields, and sucrose percentages for the July harvest are in Table 9.4. In comparing the data in Tables 9.1 and 9.4, several important differences are apparent. While the nitrogen treatments are identical, substantially more water was applied for generating yields for the July harvest made about two months after the early commercial May harvest. On the average, 18.0, 14.5, 11.7, 15.0, and 12.0 more acre-inches were applied with $I_1, \ldots, I_5$, respectively, for the late harvest over the early harvest. The additional water and the longer growing season resulted in considerably higher average root yields in the late harvest. Sucrose percentages also increased.

The most striking difference between the two harvests is the drop in average top yields, especially at higher levels of fertilization. With the longer growing season, a shift in plant use of available water and soil nutrients from top growth to root growth and sucrose content occurred. In addition, top growth at this site normally declines after reaching a peak in May or early June. Plants begin to shed leaves in late June and July. Sucrose percentages also tend to be somewhat higher for the late as compared to the early harvest.

Root yield. Average root and top yields are also summarized in Table 9.4. When average yields associated with $N = 360$ pounds per acre are compared to those for $N = 180$, average yields for roots are decreasing at all levels of water applied while those for tops are higher. Based on analyses of variance, no block effect was observed among either the root or top yield data.

As with data for the May harvest, the square root function in (9.3) provided a good fit for these data. Most estimated coefficients are statistically significant at conventional levels. Since the $LOFF$ is not significant at the 5 percent or higher level, (9.3) is used to derive additional information about the physical and economic relationships implicit in this experiment. Predicted yields derived from (9.3) are listed in Table 9.5. The maximum yield is in the vicinity of 33.5 tons per acre.

$$\text{yield} = -116.6879^{**} - 3.2661^{**}W_1 - .1455^{***}N_1 + 38.6938^{**}W_1^{.5}$$
$$\text{(roots)} \quad (53.87) \qquad (1.588) \qquad (.021) \qquad (18.60)$$

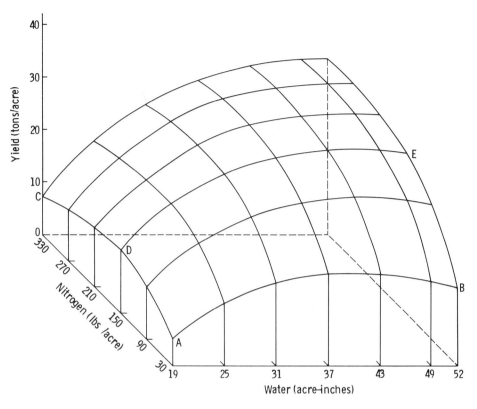

FIG. 9.1. Production surface for roots at 15 percent sucrose estimated from (9.3) (Mesa, Arizona, 1971–72, July harvest).

$$+ 1.6053N_i^{.5} + .4568^{**}W_i^{.5}N_i^{.5} \qquad (9.3)$$
$$(1.444) \qquad (.219)$$

$$R^2 = .777 \qquad F = 26.42^{***} \qquad LOFF = 2.060$$

Several other relationships are derivable from (9.3). A three-dimensional yield response surface to water and nitrogen is plotted in Figure 9.1. When sufficiently large quantities of water and (or) nitrogen are applied, yields begin decreasing. Curve AB, for example, in Figure 9.1, represents a yield response curve to water when nitrogen is applied at the rate of 30 pounds per acre. The yield response curve reaches a maximum when about 40 acre-inches of water are applied. When nitrogen is applied at a rate of 150 pounds per acre, represented by curve DE in Figure 9.1, yield reaches a maximum when water applied equals about 46 acre-inches. Curve ADC, on the other hand,

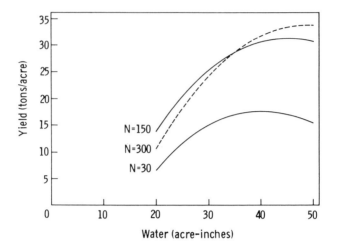

FIG. 9.2. Yield response curves to water for roots at 15
          percent sucrose derived from (9.3) assuming
          specified levels of nitrogen (Mesa, Arizona,
          1971–72, July harvest).

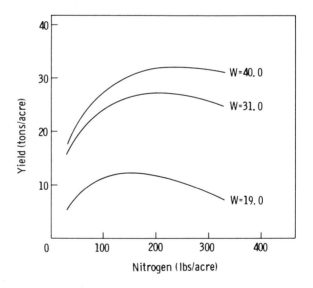

FIG. 9.3. Yield response curves to nitrogen for roots at 15
          percent sucrose derived from (9.3) assuming
          specified levels of water (Mesa, Arizona,
          1971–72, July harvest).

depicts the yield response to successively larger applications of nitrogen when water availability is fixed at 19 acre-inches.

Yield response curves to water and nitrogen are drawn in Figures 9.2 and 9.3, respectively. These curves are abstractions from Figure 9.1. Based on Figure 9.2, fertilizer should never be applied at a rate of 300 pounds per acre when the use-level for water is less than 35 acre-inches. Predicted yields within this range of water and nitrogen are always less than when the level of nitrogen is 150 pounds per acre. For the three levels of fertilization, $N = 30$, $N = 150$, and $N = 300$ pounds per acre, estimated yield reaches a maximum in the vicinities of 40, 46, and 51 acre-inches of water, respectively.

The yield response curves of nitrogen in Figure 9.3 are such that the initial response to fertilizer is fairly strong but then quickly levels off and eventually declines. When water is applied at a rate of 19 acre-inches, for example, yield reaches a maximum when about 150 pounds of nitrogen per acre are applied. When 40 acre-inches of water are applied, yield is maximized when fertilizer is approximately 240 pounds per acre.

The marginal product curves for water, nitrogen, and land as derived from (9.3) are plotted in Figures 9.4-9.6, respectively. All curves are de-

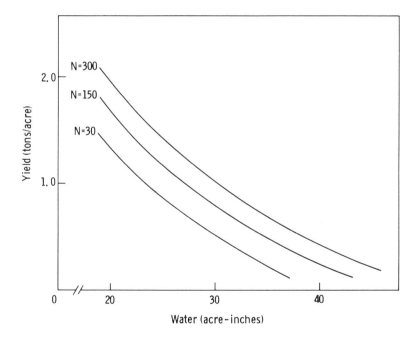

FIG. 9.4. Marginal product curves for water in production of roots at 15 percent sucrose at specified levels of nitrogen (Mesa, Arizona, 1971-72, July harvest).

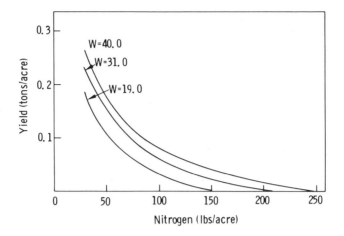

FIG. 9.5. Marginal product curves for nitrogen in production of roots at 15 percent sucrose at specified
levels of water (Mesa, Arizona, 1971–72, July
harvest).

clining at a relatively high rate.   These phenomena would suggest rather
narrow ranges of substitutability among these inputs.

The vertical distance between the curves reflects the increase in marginal
product obtainable by increasing the fixed input(s) to the specified level.
Referring to Figure 9.6, substantial increases in the marginal product of land
are generated when the availability levels of water and nitrogen are increased.

Isoquants for four yield levels are plotted in Figure 9.7. These isoquants
are based on (9.4) where alternative levels for yield and N_1 are assumed and
the corresponding level of W_1 derived. Subscripts for the water and nitrogen
variables have been omitted in (9.4).

$$W = [\{- (38.6938 + .4568N^{.5}) \pm [(38.6938 + .4568N^{.5})^2$$

$$+ (13.0644) (-\text{yield} - 116.6879 - .1455N + 1.6053N^{.5}]^{.5}\}/(-6.5322)]^2$$

$$(9.4)$$

Recall that the value of the slope at a point on an isoquant is equal to the
marginal rate of substitution between the two inputs at that point. Referring
to isoquant $Y = 20$ in Figure 9.7, segment AB denotes the range of substitution between W and N. That is, as successively more water is applied, lesser
amounts of nitrogen are required to produce a per acre yield of 20 tons.
Beyond point A, water and nitrogen are complementary in that successively
larger quantities are required to generate $Y = 20$.

Isoclines representing three price situations are superimposed in Figure
9.7. At points of intersection between isoquants and isoclines, the marginal

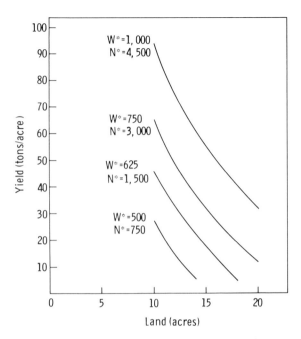

FIG. 9.6. Marginal product curves for land in the production of roots at 15 percent sucrose with specified availabilities of acre-inches of water and pounds of nitrogen (Mesa, Arizona, 1971–72, July harvest).

rate of substitution between the two inputs is equal to their inverse price ratio. The corresponding levels of water and nitrogen represent the least-cost combination of producing the specified yield when water and nitrogen are only inputs having economic values. As the price of water increases relative to the price of nitrogen, movement from isocline I_1 to I_2 to I_3 occurs. In the process, relatively more nitrogen and less water are used in producing the specified output level.

Top yield. No block effect was observed for the top yield data. Based on the *LOFF* statistics, all the commonly used polynomials give a good fit to the data. The estimated model using a square root formulation is in (9.5). None of the estimated coefficients are statistically significant at conventional levels. In addition, negative and positive coefficients would be expected for the linear and square root terms, respectively.

$$\text{yield} = 114.8708 + 2.5338W_1 + .0190N_1 - 33.8213W_i^{.5}$$
(tops)   (81.74)   (2.410)   (0.31)   (28.22)

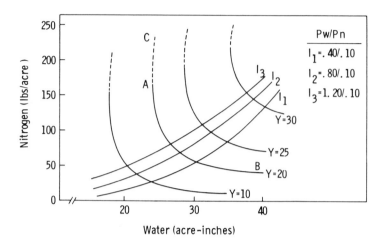

FIG. 9.7. Isoquants and isoclines for specified per acre
yields of roots at 15 percent sucrose and indicated price ratios (Mesa, Arizona, 1971-72, July
harvest).

$$- 1.3535 N_i^{.5} + .2829 W_i^{.5} N_i^{.5} \qquad\qquad (9.5)$$
$$(2.191) \qquad\quad (.332)$$
$$R^2 = .831 \qquad F = 37.45^{***} \qquad LOFF = 1.276$$

The relatively high standard errors and incorrect signs may be a consequence
of roots, tops, and sucrose percentages being joint products. That is, inter-
dependencies exist among these outputs. Predicted yields derived from (9.5)
are given in Table 9.6. Because of the unusual properties of (9.5), the yield
response to nitrogen is positive at all levels of water applied. On the other
hand, the yield response to water is negative. That is, yield declines as successively larger quantities of water are applied at all levels of nitrogen.

MESA, ARIZONA, 1970-71

This experiment is essentially identical to the one described for 1971-72.
Seed was planted September 24, 1970, and the two harvests were completed
on May 4 and July 7, 1971. Plant virus diseases occurred relatively late in the
growing season, but yields were not likely affected. Rainfall during the
growing season totaled about 0.8 inch. The level of preplant soil nitrogen in
the upper 3 feet was estimated at 40.8 pounds per acre.

May harvest

Treatment levels, root and top yields, and corresponding sucrose percentages are summarized in Table 9.7. Considerable variability exists among both
root and top yields when no nitrogen is applied.

Root yield. A block effect was operative among root yield data for this experiment. The 1.5 polynomial in (9.6) represents yield-water-nitrogen relationships where X_1 embodies the dummy variable for block effect. Except for the intercept and the interaction term, all estimated coefficients are statistically significant at the 5 percent or higher level. The low *LOFF* statistics indicate that the 1.5 polynomial is adequate or suitable for this experiment. Predicted yields derived from (9.6) are in Table 9.8.

$$\text{yield} = 19.8582 - 2.7605^{***}X_1 + 3.5809^{**}W_1 + .0736^{***}N_1$$
$$\text{(roots)} \quad (13.14) \quad\quad (.748) \quad\quad (1.532) \quad\quad (.023)$$

$$- .4655^{**}W_1^{1.5} - .0035^{***}N_1^{1.5} + .0006W_1N_1 \tag{9.6}$$
$$(.202) \quad\quad (.001) \quad\quad (.0006)$$

$$R^2 = .616 \quad F = 9.883^{***} \quad LOFF: \text{Block 1} = 1.121$$
$$\text{Block 2} = .072$$

Top yield. No block effect was observed among the top yield data. The 1.5 polynomial in (9.7) was fitted to these data. Neither of the estimated coefficients for the nitrogen variables is statistically significant at conventional levels, although signs on coefficients are those expected. This seems to contradict the strong average yield response to nitrogen at all levels of water in Table 9.7. When the average yield responses to nitrogen curves are plotted, they tend to be either linear or convex to the origin. Consequently, there is not a strong quadratic, square root, or 1.5 effect for nitrogen within these data. When these latter effects are omitted, the linear effect for nitrogen is still not statistically significant at conventional levels. The interaction term seems to be capturing much of the effect of nitrogen. When both water and nitrogen are increased, average yields nearly always increase.

$$\text{yield} = -59.9810^{**} + 8.2546^{***}W_1 + .0164N_1 - 1.0846^{***}W_1^{1.5}$$
$$\text{(tops)} \quad (22.78) \quad\quad (2.657) \quad\quad (.040) \quad\quad (.350)$$

$$- .0011N_1^{1.5} + .0034^{***}W_1N_1 \tag{9.7}$$
$$(.002) \quad\quad (.001)$$

$$R^2 = .905 \quad F = 72.03^{***} \quad LOFF = .539$$

As noted before, roots and tops are also joint products. Yields predicted from (9.7) are listed in Table 9.9.

July harvest

Treatment levels and yield data for the July harvest are given in Table 9.10. Compared with average yields for the May harvest in Table 9.7, average root yields from the July harvest are nearly doubled while average top yields are substantially lower. Based on analyses of variance, no block effect was observed for either the root or top yield data.

Root yield. The quadratic function fitted to the root yield data adjusted to a 15 percent sucrose level is given in (9.8). The estimated coefficients for the nitrogen variables are highly significant. This is consistent with the average yield response to nitrogen implicit in Table 9.10. The yield response to water is not as strong, especially at low levels of nitrogen. Since the *LOFF* statistic is low and the other test statistics and coefficient signs are reasonably good, (9.8) is used in deriving the predicted yields in Table 9.11.

$$\text{yield} = 6.6485 + .7830\overset{*}{W_1} + .0702\overset{***}{N_1} - .0109W_1^2$$
$$(\text{roots}) \quad (8.588) \quad (.463) \quad (.025) \quad (.007)$$

$$- .00019\overset{***}{N_1^2} + .00095W_1N_1 \tag{9.8}$$
$$(.00004) \quad (.0007)$$

$$R^2 = .616 \quad F = 12.21\overset{***}{} \quad LOFF = .594$$

The water variable can also be defined in terms of total water available for plant growth. Accordingly, estimates of W_2 are given in Table 9.10. The quadratic model incorporating W_2 is given in (9.9). The properties of (9.9) are reasonably similar to those for (9.8).

$$\text{yield} = 7.9642 + .8094W_2 + .0652\overset{**}{N_1} - .0127W_2^2$$
$$(\text{roots}) \quad (8.887) \quad (.551) \quad (.028) \quad (.009)$$

$$- .0002\overset{***}{N_1^2} + .0012W_2N_1 \tag{9.9}$$
$$(.00004) \quad (.0008)$$

$$R^2 = .606 \quad F = 11.71\overset{***}{} \quad LOFF = .724$$

Top yield. The 1.5 polynomial in (9.10) embodies the top yield-water-nitrogen relationships for this experiment. There was no block effect among these data. Based on the yield data in Table 9.10, the average yield response to nitrogen is strong at all levels of water treatments. Since the coefficients estimated for the nitrogen variables in (9.10) are not statistically significant at conventional levels, the yield response to nitrogen is apparently embodied in the water × nitrogen interaction term. Predicted yields derived from (9.10) are in Table 9.12.

$$\text{yield} = -15.4530\overset{*}{} + 1.7444\overset{**}{W_1} - .0057N_1 - .1941\overset{**}{W_1^{1.5}}$$
$$(\text{tops}) \quad (9.056) \quad (.765) \quad (.026) \quad (.087)$$

$$- .0009N_1^{1.5} + .0020\overset{***}{W_1N_1} \tag{9.10}$$
$$(.001) \quad (.0005)$$

$$R^2 = .926 \quad F = 94.95\overset{***}{} \quad LOFF = 1.737$$

When the water variable is defined as W_2, the square root function in (9.11) seems appropriate. In comparing (9.11) with (9.10), estimated coefficients for the nitrogen variables are relatively important in the former while the statistical significance of the water variables is higher in the latter.

$$\text{yield} = -27.2770 - 2.0036^*W_2 + .0436^*N_1 + 17.2486^*W_2^{.5}$$
$$\text{(tops)} \quad (26.57) \quad (1.013) \quad (.024) \quad (10.24)$$

$$- 4.7006^{***}N_1^{.5} + .8348^{***}W_2^{.5}N_1^{.5} \tag{9.11}$$
$$(1.124) \quad (.201)$$

$$R^2 = .923 \quad F = 91.05^{***} \quad LOFF = 1.980$$

MESA, ARIZONA, 1969-70

The 1969-70 experiment is similar to the 1970-71 and 1971-72 trials. Seed was planted September 24, 1969, and plots were harvested May 5 and July 7, 1970. The estimated level of preplant soil nitrogen (NO_3) in the upper 3 feet was 45.6 pounds per acre. Rainfall exceeding 0.25 inch at any occurrence totaled 2.45 inches during the growing season. Plots receiving no nitrogen treatment exhibited foliar symptoms of nitrogen deficiency by November 15, 1969. The relatively lower yields for the 1969-70 experiment are at least partially the result of the heavy incidence of virus yellows diseases. Beet western yellows virus was observed as early as December 1970. Plots were also later infected with beet yellows virus and mosaic virus.

May harvest

Treatment levels and yield data are given in Table 9.13. Average root yields adjusted to 15 percent sucrose and acreage top yields are also listed.

Root yield. The average yield response to water applied is weak and tends to be negative. That is, average yields at levels of nitrogen other than $N = 90$ increase as less water is applied. The yield response to nitrogen is relatively stronger. These relationships are reflected in (9.12), the 1.5 polynomial fitted to the root yield-water-nitrogen data. The estimated coefficients for the nitrogen variables have considerably higher levels of statistical significance than those estimated for the water variables. A block effect also existed for this experiment. Predicted yields derived from (9.12) are given in Table 9.14.

$$\text{yield} = 4.2615 - .8084^*X_1 + .1698W_1 + .0968^{***}N_1 - .0428W_1^{1.5}$$
$$\text{(roots)} \quad (10.95) \quad (.449) \quad (1.515) \quad (.020) \quad (.212)$$

$$- .0042^{***}N_1^{1.5} + .0004W_1N_1 \tag{9.12}$$
$$(.0006) \quad (.0006)$$

$$R^2 = .757 \quad F = 19.25^{***} \quad LOFF: \text{Block } 1 = .693$$
$$\text{Block } 2 = 1.197$$

Top yield. Referring to Table 9.13, the average yield response to nitrogen is very strong, but the response to water is inconclusive. When $N = 0$ and 180, for example, the average yield response curves for water are concave and convex to the origin, respectively. No block effect was observed among the top yield data. The square root function fitted to the data is given in (9.13). Estimated coefficients for the nitrogen variables are highly significant. The expected signs for the water variables are reversed. Predicted yields derived from (9.13) are summarized in Table 9.15.

$$
\begin{aligned}
\text{yield} &= -6.9482 - 1.9382 W_1 + .1202^{***} N_1 + 12.6135 W_1^{.5} \\
\text{(tops)} &\quad (69.22) \quad\quad (3.541) \quad\quad (.026) \quad\quad\quad (31.76)
\end{aligned}
$$

$$
\begin{aligned}
&\quad\quad - 4.4442^{**} N_1^{.5} + .6647^{*} W_1^{.5} N_1^{.5} \\
&\quad\quad\quad (1.902) \quad\quad\quad (.363)
\end{aligned}
\tag{9.13}
$$

$$
R^2 = .941 \quad\quad F = 120.52^{***} \quad\quad LOFF = 2.222
$$

July harvest

Treatment levels and yield data for the July harvest are given in Table 9.16. Average root yields for the 1969–70 July harvest are considerably less than for the comparable 1971–72 and 1970–71 harvests. Levels of nitrogen for the three experiments are identical, and levels of irrigation are comparable.

Root yield. No block effect existed among the root yield data. The 1.5 polynomial in (9.14) represents root yield-water-nitrogen relationships for the July harvest. Only the estimated coefficients for the nitrogen variables and interaction term are statistically significant at the 10 percent or higher level. When the water variable is defined in terms of total water available, W_2, the quadratic form in (9.15) is estimated.

$$
\begin{aligned}
\text{yield} &= -15.8179 + 1.7595 W_1 + .1337^{***} N_1 - .1979 W_1^{1.5} \\
\text{(roots)} &\quad (15.92) \quad\quad (1.304) \quad\quad (.028) \quad\quad\quad (.142)
\end{aligned}
$$

$$
\begin{aligned}
&\quad\quad - .0072^{***} N_1^{1.5} + .0011^{**} W_1 N_1 \\
&\quad\quad\quad (.001) \quad\quad\quad (.0005)
\end{aligned}
\tag{9.14}
$$

$$
R^2 = .776 \quad\quad F = 26.39^{***} \quad\quad LOFF = .824
$$

In comparing (9.15) with (9.14), however, test statistics for the two models are not substantially different. Predicted yields derived from (9.14) are given in Table 9.17.

$$
\begin{aligned}
\text{yield} &= -7.0589 + .9862 W_2 + .0450^{*} N_1 - .0170^{*} W_2^2 \\
\text{(roots)} &\quad (9.029) \quad\quad (.610) \quad\quad (.026) \quad\quad (.010)
\end{aligned}
$$

$$
\begin{aligned}
&\quad\quad - .0002^{***} N_1^2 + .0018^{**} W_2 N_1 \\
&\quad\quad\quad (.00003) \quad\quad\quad (.0007)
\end{aligned}
\tag{9.15}
$$

$$
R^2 = .778 \quad\quad F = 26.59^{***} \quad\quad LOFF = .794
$$

Top yield. No block effect also existed for the top yield data. Based on the square root function in (9.16), most of the nitrogen and water treatment effect on yield is captured by the interaction term. The estimated standard errors for the other coefficients are relatively high.

$$\text{yield} = -38.2263 - 1.5412 W_1 - .0017 N_1 + 15.8143 W_1^{.5}$$
$$\text{(tops)} \quad (53.16) \qquad (1.592) \qquad (.023) \qquad (18.54)$$

$$- 2.2368^* N_1^{.5} + .4911^{**} W_1^{.5} N_1^{.5} \tag{9.16}$$
$$(1.324) \qquad (.203)$$

$$R^2 = .829 \qquad F = 36.84^{***} \qquad LOFF = 1.554$$

When the water variable is represented by W_2, the square root function is reestimated as (9.17). Predicted yields derived from (9.16) are listed in Table 9.18.

$$\text{yield} = -5.7127 - .7742 W_2 - .0144 N_1 + 5.3629 W_2^{.5}$$
$$\text{(tops)} \quad (41.33) \qquad (1.577) \qquad (.024) \qquad (16.33)$$

$$- 1.8474 N_1^{.5} + .5017^* W_2^{.5} N_1^{.5} \tag{9.17}$$
$$(1.459) \qquad (.252)$$

$$R^2 = .816 \qquad F = 33.80^{***} \qquad LOFF = 1.993$$

Mesa aggregated (May harvest)
Since the experimental design for the three years is similar, the annual data can be aggregated into a composite function representing an average across the three years. Two aggregation procedures are considered. The first is to aggregate using a weather variable. The implicit assumption is that interyear variation in yield not attributable to differences in water and nitrogen treatment levels is due to differences in climatic conditions. Recall that rainfall has been previously added to the irrigation treatment levels. The second approach is to use dummy variables for reflecting any year-to-year variation in soil conditions, management, and environmental factors. The dummy variable procedure provides a more comprehensive representation of interyear variation. On the other hand, specific sources of variation are not identifiable.

1969-70 to 1971-72 combined
Class A pan evaporation (PE) data are used as a proxy for the weather variable. These data are hypothesized to reflect important weather factors including temperature, humidity, and to some extent wind velocity. Let PE_1 represent the sum of daily pan evaporation readings from planting up to 2-3 weeks prior to the May harvest. These are estimated in inches as

1971-72 37.9

1970-71 40.3

1969-70 39.4

The dummy variables have been constructed as

$$X_1 \quad X_2$$

	X_1	X_2
1971–72	0	0
1970–71	1	0
1969–70	0	1

With this formulation, X_1 and X_2 are used to estimate whether the yearly effects of 1970-71 and 1969-70 are statistically significantly different from those for 1971-72 where the latter is arbitrarily chosen as the base year.

Root yield. When the data are aggregated with a weather variable, the 1.5 polynomial in (9.18) is used to represent the yield-water-nitrogen relationships for roots adjusted to 15 percent sucrose. These estimated coefficients for the nitrogen variables are highly significant; those for the water variables are significant only at lower levels but have appropriate signs. This is consistent with earlier results in that the estimated water coefficients in the individual year functions were statistically significant at conventional levels for only the 1970-71 function.

$$\text{yield} = 69.9089^{***} + .5820W_1 + .1001^{***}N_1 - .0466W_1^{1.5}$$
$$(\text{roots}) \quad (19.41) \qquad (1.428) \qquad (.022) \qquad (.192)$$

$$- .0040^{***}N_1^{1.5} - .0001W_1N_1 - 1.7923^{***}PE_1 \qquad \textbf{(9.18)}$$
$$(.001) \qquad\quad (.0006) \qquad\quad (.347)$$

$$R^2 = .438 \qquad F = 16.24^{***}$$

The negative coefficient for PE_1 in (9.17) implies an inverse relationship between weather, as defined, and yield. That is, when high temperatures and (or) low humidity, for example, cause the level of pan evaporation to increase, yields are adversely affected and vice versa. Also, the negative coefficient for the interaction term indicates that water and nitrogen are substitutes in the production of roots.

When dummy variables are incorporated, the 1.5 polynomial in (9.19) is estimated. The magnitudes of estimated coefficients and their levels of statistical significance are similar to those for (9.18). Since the estimated coefficients for the dummy variables are highly significant, the composite effect of environmental and managerial factors for both 1970-71 and 1969-70 differ significantly from 1971-72, the assumed base year. Individual year functions can be derived from (9.19) by substituting in appropriate values for X_1 and X_2.

$$\text{yield} = 6.3477 - 9.0604^{***}X_1 - 3.5388^{***}X_2 + .6451W_1 + .1068^{***}N_1$$
$$(\text{roots}) \quad (6.659) \qquad (.496) \qquad\quad (.481) \qquad\quad (.815) \qquad (.013)$$

$$- .0826W_1^{1.5} - .0045^{***}N_1^{1.5} + .0001W_1N_1 \qquad \textbf{(9.19)}$$
$$(.110) \qquad\quad (.0005) \qquad\quad (.0003)$$

$$R^2 = .818 \qquad F = 79.85^{***}$$

Top yield. Aggregated functions were also fitted to the top yield data. The square root function incorporating a weather variable (PE_1) is given in (9.20). The estimated coefficient for the weather variable is not statistically significant at conventional levels. The square root function incorporating dummy variables X_1 and X_2 is in (9.21). Since the estimated coefficient for X_1 is highly significant, the experimental factors other than W_1 and N_1 for the 1970-71 test were significantly different from those for the 1971-72 experiment. This conclusion does not apply to a comparison of the 1969-70 versus 1971-72 test conditions.

$$\text{yield} = 7.5889 - 1.0302W_1 + .0694^{***}N_1 + 8.5009W_1^{.5}$$
$$(\text{tops}) \quad (60.30) \quad (2.409) \quad (.029) \quad (22.63)$$

$$- 3.1100^{***}N_1^{.5} + .6360^{***}W_1^{.5}N_1^{.5} - .4426PE_1 \qquad (9.20)$$
$$(1.403) \quad (.281) \quad (.548)$$

$$R^2 = .819 \quad F = 94.40^{***}$$

$$\text{yield} = -1.6952 - 8.2479^{***}X_1 - .1373X_2 - 1.3418W_1$$
$$(\text{tops}) \quad (41.89) \quad (1.068) \quad (1.036) \quad (1.889)$$

$$+ .0605^{***}N_1 + 8.6879W_1^{.5} - 3.1990^{***}N_1^{.5}$$
$$(.023) \quad (17.75) \quad (1.100)$$

$$+ .7109^{***}W_1^{.5}N_1^{.5} \qquad (9.21)$$
$$(.220)$$

$$R^2 = .890 \quad F = 142.96^{***}$$

1970-71 and 1971-72 combined

Recall that observed yields for the 1969-70 experiment were relatively low partially due to the incidence of virus yellows diseases. Omitting these data, the combined functions for roots and tops are in (9.22) and (9.23), respectively. The aggregated functions incorporating the weather variable are not listed. The dummy variables are formulated as

$$X_1$$

1971-72	0
1970-71	1

In comparing (9.22) with (9.19), the R^2 and the F values for the two-year combined function are somewhat lower, but a larger number of the estimated coefficients are statistically significant at conventional levels. For the top yield data, the importance of statistical significance of the estimated coefficients has shifted from nitrogen in (9.21) to water in (9.23). Predicted yields derived from (9.22) and (9.23) are listed in Tables 9.19 and 9.20, respectively.

$$\text{yield} = 63.0798^{**} - 2.9763^{***}X_1 - 2.4426^{**}W_1 - .0834^{***}N_1$$
$$(\text{roots}) \quad (29.45) \quad\quad (.542) \quad\quad (1.186) \quad\quad (.014)$$

$$+ 25.2549^{**}W_1^{.5} + 2.0825^{***}N_1^{.5} + .0926W_1^{.5}N_1^{.5} \qquad (9.22)$$
$$(11.72) \quad\quad (.639) \quad\quad (.129)$$

$$R^2 = .637 \quad F = 23.68^{***}$$

$$\text{yield} = -39.0658^{*} + .4515X_1 + 5.1990^{**}W_1 + .0556N_1$$
$$(\text{tops}) \quad (21.40) \quad (1.161) \quad (2.437) \quad (.036)$$

$$- .6618^{**}W_1^{1.5} - .0024N_1^{1.5} + .0027^{***}W_1N_1 \qquad (9.23)$$
$$(.318) \quad\quad (.002) \quad\quad (.001)$$

$$R^2 = .862 \quad F = 84.45^{***}$$

Mesa aggregated (July harvest)

The procedures described for aggregating the May harvest data also apply here. Results from combining three years of data are presented first and then data for the 1969–70 test are later omitted. Only those functions incorporating dummy variables are specified here.

1969–70 to 1971–72 combined

The square root function in (9.24) was selected to represent yield-water-nitrogen relationships for roots adjusted to 15 percent sucrose.

$$\text{yield} = -41.7163^{**} - 11.3857^{***}X_1 + 3.5912^{***}X_2 - 1.6238^{***}W_1$$
$$(\text{roots}) \quad (20.18) \quad\quad (.629) \quad\quad (.665) \quad\quad (.607)$$

$$- .1261^{***}N_1 + 16.8654^{**}W_1^{.5} + 1.6267^{***}N_1^{.5}$$
$$(.014) \quad\quad (6.896) \quad\quad (.624) \qquad\qquad (9.24)$$

$$+ .3757^{***}W_1^{.5}N_1^{.5}$$
$$(.105)$$

$$R^2 = .884 \quad F = 134.42^{***}$$

All estimated coefficients are highly significant. Consequently, an interyear effect on yield other than that generated by water and nitrogen treatments is observed. The aggregated function for the top yield data is given in (9.25). Compared with (9.24), relatively few of the estimated coefficients for the treatment variables are significant at the 10 percent or higher level.

$$\text{yield} = -6.3199 - 1.6398^{**}X_1 + 5.1320^{***}X_2 - .5891W_1$$
$$(\text{tops}) \quad (24.11) \quad (.752) \quad\quad (.795) \quad\quad (.725)$$

$$+ .0107N_1 + 4.6518W_i'^5 - 2.1984^{***}N_i^5$$
$$(.016) \qquad (8.241) \qquad (.746) \tag{9.25}$$

$$+ .4721^{***}W_i'^5N_i^5$$
$$(.125)$$

$$R^2 = .842 \qquad F = 94.75^{***}$$

1970-71 and 1971-72 combined

After omitting data from the 1969-70 test, the aggregated function for roots at 15 percent sucrose is given in (9.26). All estimated coefficients are statistically significant at the 10 percent or higher level.

$$\text{yield} = -40.6898^{*} + 3.5703^{***}X_1 - 1.4859^{**}W_1 - .1273^{***}N_1$$
$$\text{(roots)} \quad (23.93) \qquad (.732) \qquad (.730) \qquad (.081)$$

$$+ 15.8137^{*}W_i'^5 + 1.9322^{**}N_i^5 + .3316^{**}W_i'^5N_i^5 \tag{9.26}$$
$$(8.143) \qquad (.814) \qquad (.143)$$

$$R^2 = .654 \qquad F = 25.54^{***}$$

The quadratic function in (9.27) was fitted to the top yield data. Most of the estimated coefficients for the treatment variables have relatively high standard errors. Predicted yields derived from (9.26) and (9.27) are listed in Tables 9.21 and 9.22, respectively.

$$\text{yield} = -6.6950 + 5.2495^{***}X_1 + .3741W_1 + .0089N_1$$
$$\text{(tops)} \quad (7.252) \qquad (.810) \qquad (.370) \qquad (.017)$$

$$- .0041W_i^2 - .00004^{*}N_i^2 + .0014^{***}W_1N_1 \tag{9.27}$$
$$(.005) \qquad (.00003) \qquad (.0005)$$

$$R^2 = .862 \qquad F = 84.17^{***}$$

YUMA MESA, ARIZONA, 1970-71

Spreckles 301-H was planted October 6, 1970, in 20-inch rows on Superstition Fine Sand. Prior to planting, 200 pounds of P_2O_5 per acre and 25 pounds of nitrogen per acre were uniformly applied. Plants were thinned to an average within-row spacing of 6 inches. Irrigation treatments $I_1, \ldots, I_5$ were applied when the percent available water used dropped to an average of 85, 75, 65, 55, and 45 percent, respectively. Initially, irrigations were made when the soil moisture reached the specified level in the 12-24-inch zone. As more extensive root systems developed, average soil moisture throughout the 12-36-inch depth was used to determine irrigation timing. Soil at this site has an estimated available water-holding capacity of only 3.1 acre-inches in the top 4 feet. Rainfall was not a factor in this experiment.

Slow plant growth and foliar symptoms indicated that the nitrogen was being leached as water was applied for initial growth and establishment of plant stands. About 10 irrigations of applications of 3 inches were made during this period. Soil at this site has a low base exchange capacity. Adding nitrogen to the water stimulated plant growth. The total nitrogen applied ranged from 165 to 565 pounds per acre with treatment levels designated by 100 pound per acre increments. Plots were harvested June 20, 1971.

Treatment levels together with corresponding root, top, and sucrose yields are summarized in Table 9.23. Both water and nitrogen treatment levels are relatively high. Average yields for roots adjusted to 15 percent sucrose and for tops are also listed.

Root yield. The average yield response to water is not strong for this experiment. When $N = 165$ pounds per acre, average yield declines with successively larger applications of water. The initial average yield response to nitrogen is especially pronounced for I_1, I_3, and I_5 water treatments. At heavier levels of nitrogen, however, average yields decline.

Based on an analysis of variance, a block effect existed for this experiment. Among the commonly used polynomials, the 1.5 polynomial in (9.28) was selected for quantifying yield-water-nitrogen relationships.

$$
\begin{aligned}
\text{yield} &= -44.45\overset{**}{3}5 - 2.07\overset{***}{2}0X_1 + .7507W + .39\overset{***}{0}9N \\
\text{(roots)} \quad &(20.26) \quad\quad (.778) \quad\quad (.782) \quad\quad (.033) \\[2mm]
&\quad - .0682W^{1.5} - .01\overset{***}{4}6N^{1.5} + .000\overset{***}{7}5WN \\
&\quad\;\; (.059) \quad\quad\;\; (.001) \quad\quad\quad (.0002)
\end{aligned}
\tag{9.28}
$$

$$
R^2 = .926 \quad\quad F = 77.\overset{***}{3}8 \quad\quad LOFF\text{: Block 1} = 2.957
$$
$$
\text{Block 2} = 2.422
$$

Dummy variable X_1 is assigned a zero value for Block 1 and a value of one for Block 2. Estimated coefficients for the water variables are not statistically significant at 10 percent levels although signs are appropriate. This is consistent with the average yield response to water implicit in Table 9.23. The other estimated coefficients are significant at the 5 percent or higher level. Predicted yields derived from (9.28) are listed in Table 9.24.

Top yield. There was no block effect among the sugar beet top yield data. The quadratic function fitted to the data is in (9.29).

$$
\begin{aligned}
\text{yield} &= -12.4896 + .4733W - .065\overset{*}{3}N - .0047W^2 \\
\text{(tops)} \quad &(24.72) \quad\quad (.627) \quad\quad (.034) \quad\quad (.004) \\[2mm]
&\quad + .00003N^2 + .00\overset{***}{1}3WN \\
&\quad\;\; (.00003) \quad\quad (.0003)
\end{aligned}
\tag{9.29}
$$

$$
R^2 = .899 \quad\quad F = 67.\overset{***}{7}0 \quad\quad LOFF = .718
$$

Predicted yields derived from (9.29) are given in Table 9.25.

YUMA MESA, ARIZONA, 1969-70

Spreckles 301-H was planted November 12, 1969, in two rows spaced 12 inches apart per 40-inch bed at the Yuma Mesa site. Plants were later thinned to 6-inch intervals.

About 33 acre-inches of water were applied in six uniform irrigations from the time of planting through January for initial plant growth and stand establishment. Variable irrigation treatments $I_1, \ldots, I_5$ were made when the percent of available soil moisture fell to 4.5, 5.6, 6.7, 5.1, and 6.2 percent, respectively. Because of the sandy nature of this soil and its low available water-holding capacity, the number of irrigations ranged from 16 for treatment combination I_1N_1 to 32 for I_5N_5. Total water applied ranged from 56.3 to 72.5 acre-inches. Rainfall exceeding 0.25 inch at any occurrence totaled about 2.7 inches for the growing season.

Side-dress applications of 100 pounds of P_2O_5 per acre and 15 pounds of ammonium nitrate per acre were applied after planting. As with the 1970-71 experiment, the lack of plant response to side-dress application necessitated adding subsequent applications of N to the irrigation water. This appeared to be the most effective method of getting adequate nitrogen into the root zone. The total quantity of nitrogen applied ranged from 86 pounds per acre for treatment level N_1 to 386 pounds per acre for N_5. Plots were harvested July 1-2, 1970.

Treatment levels and respective root, top, and sucrose yields are listed in Table 9.26. Sucrose percentages are available only for roots harvested from Blocks 1 and 2.

Root yield. The yield response to water in Table 9.26 is strong only at the highest level of nitrogen. A response to nitrogen is observable at all levels of water application. The quadratic function in (9.30) was selected to quantify yield-water-nitrogen relationships for the root yields. Equation (9.30) is in terms of roots adjusted to 15 percent sucrose and consequently is based on yield data from Blocks 1 and 2 only. There was no block effect among these data. Both coefficients estimated for the nitrogen variables were negative. The coefficient estimated for N^2 is relatively small. Estimates of W_2 are given in Table 9.26. The quadratic formulation incorporating W_2 is given in (9.31).

$$\text{yield} = -271.5556^* + 8.9939^{**}W_1 - .2447^{**}N - .0711^*W_1^2$$
$$\text{(roots)} \quad (140.00) \qquad (4.494) \qquad (.114) \qquad (.036)$$

$$- .0001N^2 + .0045^{**}W_1N \qquad\qquad (9.30)$$
$$(.00006) \qquad (.002)$$

$$R^2 = .700 \qquad F = 17.77^{***} \qquad LOFF = .712$$

In comparison with (9.30), fewer of the estimated coefficients are statistically significant at the 10 percent level. Predicted yields derived from (9.30) are listed in Table 9.27.

$$\text{yield} = -314.634\overset{*}{0} + 9.994\overset{*}{5}W_2 - .1974N - .076\overset{*}{0}W_2^2$$
$$\text{(roots)} \quad (159.97) \quad (5.133) \quad (.133) \quad (.041)$$

$$- .00006N^2 + .0036W_2N \tag{9.31}$$
$$(.0001) \quad (.002)$$

$$R^2 = .690 \quad F = 16.\overset{***}{89}$$

Top yield. Based on an analysis of variance, the experimental conditions for Block 2 were statistically significantly different at the 5 percent level from those for Block 3. The dummy variables were set up as

$$X_1 \quad X_2$$

Block	X_1	X_2
1	1	0
2	0	1
3	0	0

The quadratic function fitted to the top yield data is specified in (9.32). Most estimated coefficients are not statistically significant at conventional levels. When the water variable is defined as W_2, the quadratic model is reformulated as (9.33). The properties of (9.32) and (9.33) are essentially the same. Predicted values derived from (9.32) are given in Table 9.28.

$$\text{yield} = 68.2034 - .6032X_1 - 1.803\overset{**}{2}X_2 - 1.8290W_1$$
$$\text{(tops)} \quad (97.73) \quad (.727) \quad (.727) \quad (3.134)$$

$$- .1106N + .0117W_1^2 - .000\overset{**}{1}N^2 + .002\overset{**}{9}W_1N$$
$$(.079) \quad (.025) \quad (.00004) \quad (.001) \tag{9.32}$$

$$R^2 = .815 \quad F = 36.\overset{***}{55} \quad LOFF: \text{Block 1} = 2.498$$
$$\text{Block 2} = .835$$
$$\text{Block 3} = 1.500$$

$$\text{yield} = 93.5484 - .5818X_1 - 1.781\overset{**}{8}X_2 - 2.6487W_2 - .0989N$$
$$\text{(tops)} \quad (90.25) \quad (.720) \quad (.720) \quad (2.896) \quad (.075)$$

$$+ .0183W_2^2 - .000\overset{**}{1}N^2 + .002\overset{**}{7}W_2N \tag{9.33}$$
$$(.023) \quad (.00005) \quad (.001)$$

$$R^2 = .819 \quad F = 37.\overset{***}{51}$$

YUMA VALLEY, ARIZONA, 1970-71

Spreckles 301-H was planted November 9, 1970, in two 12-inch rows per 40-inch bed on Glendale Silty Clay Loam. Prior to planting, 100 pounds of P_2O_5 per acre were uniformly applied. Plants were later thinned to 8-inch intervals. Irrigation treatments $I_1, \ldots, I_5$ were designed to irrigate whenever the percent of available water used in the upper 3 feet of soil dropped to an average of 85, 75, 65, 55, and 45 percent, respectively. Soil at this site has an estimated available water-holding capacity of 9.9 acre-inches in the top 4 feet. No significant amount of rainfall was recorded during the growing season. An estimated 60 pounds of preplant soil nitrogen per acre were available for plant growth. Plots were harvested June 21, 1971.

Treatment levels and corresponding yields are summarized in Table 9.29. Root and top yields are highly variable for most treatment combinations. Average yields of roots adjusted 15 percent sucrose and average top yields are also listed. No block effect was observed among either the root or top yield data.

Root yield. As pointed out earlier, individual plot yields listed in Table 9.29 are highly variable for most treatment combinations. The 1.5 polynomial in (9.34) provides a good fit to the root yield data.

$$
\begin{aligned}
\text{yield} &= -196.3593^{**} + 15.8109^{**}W + .1618^{**}N_1 - 1.6616^{**}W^{1.5} \\
\text{(roots)} \quad &\quad (83.63) \qquad (6.425) \qquad (.071) \qquad (.685) \\
&\quad - .0120^{***}N_1^{1.5} + .0025WN_1 \\
&\quad \quad (.003) \qquad (.002)
\end{aligned}
$$

$$R^2 = .481 \qquad F = 7.05^{**} \qquad LOFF = 1.156 \tag{9.34}$$

All estimated coefficients other than for the interaction term are highly significant. When the water variable is redefined as total water available, W_2, the 1.5 polynomial is reestimated, as in (9.35). The properties of (9.35) are similar to those for (9.34). Predicted yields derived from (9.34) are listed in Table 9.30.

$$
\begin{aligned}
\text{yield} &= -220.9399^{**} + 18.5329^{***}W_2 + .1040N_1 - 1.9937^{**}W_2^{1.5} \\
\text{(roots)} \quad &\quad (84.93) \qquad (6.866) \qquad (.081) \qquad (.753) \\
&\quad - .0117^{***}N_1^{1.5} + .0038W_2N_1 \\
&\quad \quad (.003) \qquad (.002)
\end{aligned}
$$

$$R^2 = .494 \qquad F = 7.41^{**} \qquad LOFF = 1.021 \tag{9.35}$$

Top yield. The quadratic function in (9.36) was selected to represent top yield-water-nitrogen relationships for this experiment. Few of the estimated coefficients are statistically significant at conventional levels. When the water variable is defined as W_2, the quadratic function is reestimated as in (9.37). Predicted yields derived from (9.36) are given in Table 9.31.

$$
\begin{aligned}
\text{yield} &= -60.2359 + 3.7204W - .2016^{*}N_1 - .0481W^2 \\
\text{(tops)} \quad &\quad (98.28) \qquad (5.021) \qquad (.103) \qquad (.064) \\
&\quad + .0001N_1^2 + .0046^{*}WN_1 \\
&\quad \quad (.0001) \qquad (.003)
\end{aligned}
$$

$$R^2 = .563 \qquad F = 9.79^{***} \qquad LOFF = .778 \tag{9.36}$$

$$\text{yield} = -46.3361 + 3.0897W_2 - .2187^*N_1 - .0410W_2^2$$
$$(\text{tops}) \quad (99.96) \qquad (5.359) \qquad (.127) \qquad (.072)$$

$$+ .0001N_1^2 + .0049W_2N_1 \tag{9.37}$$
$$(.0001) \qquad (.004)$$

$$R^2 = .571 \qquad F = 10.10^{***} \qquad LOFF = .691$$

YUMA VALLEY, ARIZONA, 1969-70

Since this experiment is similar to the 1970-71 test, only the important differences will be pointed out. The seed was planted October 15, 1969, plants were later thinned to 6-inch intervals, and plots were harvested June 29-30, 1970. Rainfall exceeding 0.25 inch at any occurrence totaled 2.7 inches during the growing season. The estimated level of preplant soil nitrogen was 25.5 pounds per acre. The experimental design for this test incorporated three blocks with each containing 22 experimental plots. Sucrose percentages were not determined for roots harvested from Block 3.

Treatment levels and corresponding root, top, and sucrose yields are given in Table 9.32. No block effect was observed among either the roots or top yield data.

Root yield. For comparability with previous functions, the quadratic function in (9.38) was estimated for root yields adjusted to 15 percent sucrose harvested from Blocks 1 and 2. The estimated coefficients for the fertilizer variables are the only ones statistically significant at the 10 percent or higher level. The quadratic form incorporating W_2 is given in (9.39). The properties of (9.38) and (9.39) are similar. Predicted yields derived from (9.38) appear in Table 9.33.

$$\text{yield} = 3.9292 + .5022W_1 + .1101^{***}N_1 - .0040W_1^2 - .0002^{***}N_1^2 - .0001W_1N_1$$
$$(\text{roots}) \quad (8.496) \quad (.623) \qquad (.033) \qquad (.011) \qquad (.00004) \qquad (.0001)$$

$$R^2 = .772 \qquad F = 25.74^{***} \qquad LOFF = .242 \tag{9.38}$$

$$\text{yield} = 3.7436 + .5151W_2 + .1121^{***}N_1 - .0038W_2^2$$
$$(\text{roots}) \quad (9.374) \quad (.719) \qquad (.036) \qquad (.013)$$

$$- .0002^{***}N_1^2 - .0002W_2N_1 \tag{9.39}$$
$$(.00005) \qquad (.001)$$

$$R^2 = .772 \qquad F = 25.67^{***} \qquad LOFF = .252$$

Top yield. The square root function in (9.40) was fitted to top yield data from all three blocks. When the water variable is defined as W_2, (9.41) is generated. Predicted yields derived from (9.40) are given in Table 9.34.

$$\text{yield} = -15.2671 - 1.0508W_1 + .0669^{***}N_1 + 10.2219W_1^{.5}$$
$$(\text{tops}) \quad (33.49) \qquad (1.417) \qquad (.023) \qquad (14.04)$$

$$- 3.9837^{***}N_i^{.5} + .5615^{**}W_i^{.5}N_i^{.5} \qquad (9.40)$$
$$(1.461) \qquad (.238)$$

$$R^2 = .816 \qquad F = 52.25^{***} \qquad LOFF = 2.297$$

$$\text{yield} = -13.2681 - .9542W_2 + .0580^{**}N_1 + 9.1705W_2^{.5}$$
$$\text{(tops)} \quad (37.60) \quad (1.676) \quad (.023) \quad (16.11)$$

$$- 3.7755^{**}N_i^{.5} + .5654^{**}W_2^{.5}N_i^{.5} \qquad (9.41)$$
$$(1.610) \qquad (.285)$$

$$R^2 = .819 \qquad F = 54.47^{***} \qquad LOFF = 2.116$$

SAFFORD, ARIZONA, 1970

Variety USH9B1 was planted February 14 on Pima Clay Loam Variant at the Safford Branch Experiment Station. The soils and groundwaters in this area are characterized by relatively high levels of salt, particularly sodium.

Irrigation treatments $I_1, \ldots, I_5$ were designed to irrigate when available soil moisture in the upper 2 feet was depleted to an average of 85, 75, 65, 55, and 45 percent, respectively. Soil at this site has an estimated available water-holding capacity of 7.5 acre-inches in the top 4 feet. Rainfall exceeding 0.25 inch at any occurrence totaled 4.2 inches during the growing season.

The five nitrogen treatments were evenly spaced at intervals of 60 pounds per acre and ranged from $N_1 = 0$ to $N_5 = 240$ pounds per acre. The level of preplant soil nitrogen was estimated as 200 pounds per acre. Above normal temperatures during July and August retarded plant growth. Some plant disease was observed that adversely affected subsequent yields. Plots were harvested November 4.

Treatment levels, root yields, and sucrose percentages are listed in Table 9.35. The quantities of water applied vary with each plot. For example, water applied for treatment combination I_1N_1 ranges from 37.2 to 46.5 acre-inches. Recall that I_1 reflects irrigating whenever available soil moisture used falls to an average of 85 percent.

No block effect was determined for this experiment. The quadratic function in (9.42) was fitted to the yield-water-nitrogen data. Since the water treatments were not replicated, the $LOFF$ statistic cannot be derived. The estimated coefficients for the water variables are highly significant. Since the signs of the linear and quadratic nitrogen variables are negative and positive, respectively, predicted yields in Table 9.36 are declining at all water levels whenever the rate of nitrogen is increased.

$$\text{yield} = -32.7317 + 2.6182^{***}W_1 - .1665^{**}N_1 - .0252^{***}W_1^2$$
$$\text{(roots)} \quad (28.62) \quad (.968) \quad (.071) \quad (.009)$$

$$+ .0001N_1^2 + .0018^{**}W_1N_1 \qquad (9.42)$$
$$(.0001) \qquad (.0008)$$

$$R^2 = .495 \qquad F = 7.459^{***}$$

FT. COLLINS, COLORADO, 1969

Seed was planted April 21 in 22-inch rows on Nunn Clay Loam at the Agronomy Research Center. Soil moisture conditions were monitored by electrical resistance block readings at the 12-inch depth. Irrigation treatments $I_1, \ldots, I_5$ reflected a maximum soil water tension of 0.7, 1.0, 3.0, 6.0, and 9.0 bars, respectively, at a 12-inch soil depth. Soil at this site has an estimated available water-holding capacity of 9 acre-inches in the upper 4 feet. Rainfall exceeding 0.25 inch at any occurrence totaled 7.2 inches during the growing season. Nitrogen was applied as a side-dressing in late June. The level of preplant soil nitrogen was estimated as 75 pounds per acre. A period of snow and cold weather prior to harvest on November 13–14 had a depressing effect on yields.

Treatment levels and corresponding root yields and sucrose percentages are in Table 9.37. The yield response to water is strong at low levels of nitrogen. Yield response to nitrogen is relatively weak.

There was no block effect for this experiment. The low yield response to nitrogen is evident in the quadratic function in (9.43).

$$\text{yield} = 26.5560^{***} - 2.3297^{**}W_1 + .0021N_1 + .0913^{***}W_1^2$$
$$\text{(roots)} \quad (8.832) \qquad (1.046) \qquad (.013) \qquad (.030)$$

$$+ .00002N_1^2 - .0010^{**}W_1N_1 \qquad\qquad (9.43)$$
$$(.00002) \qquad (.0004)$$

$$R^2 = .707 \qquad F = 18.35^{***} \qquad LOFF = 1.270$$

Neither of the coefficients for the nitrogen variables is statistically significant at conventional levels. Note that the signs for the linear and quadratic water variables are opposite of the expected form. Predicted yields from (9.43) are included in Table 9.38.

WALSH, COLORADO, 1970

Seed was planted April 6 in 22-inch rows on Baca Clay Loam at the Colorado State University Research site at Walsh. Electrical resistance blocks at the 12-inch depth were used to determine soil moisture conditions. Irrigation treatments $I_1, \ldots, I_5$ were selected to reflect a planned maximum soil water tension of 0.7, 1.0, 3.0, 6.0, and 9.0 bars, respectively, at the 12-inch soil depth. Effective rainfall during the growing season totaled about 10 inches.

Nitrogen was applied as a side-dressing in early June. The level of preplant soil nitrogen was estimated as 33 pounds per acre. Plots were harvested October 24.

Yield-water-nitrogen data are listed in Table 9.39. Yields were quite variable for most treatment combinations. Percentages of sucrose are declining at higher levels of nitrogen.

The square root function fitted to these data is in (9.44). Based on an analysis of variance, there was no block effect among the data.

$$\text{yield} = -17.7373 - .2324W_1 - .0415^{**}N_1 + 7.2851W_1^{.5}$$
$$\text{(roots)} \quad (80.84) \quad (3.263) \quad (.019) \quad (32.69)$$

$$+ .4201N_1^{.5} + .1002W_1^{.5}N_1^{.5} \tag{9.44}$$
$$(.963) \qquad (.165)$$

$$R^2 = .565 \quad F = 9.868^{***} \quad LOFF = 1.30$$

Except for N_1, the standard errors of all estimated coefficients are high, but signs are generally appropriate. Predicted yields derived from (9.44) are in Table 9.40.

PLAINVIEW, TEXAS, 1971

Great Western Monogerm was planted March 31 at a rate of 2.4 pounds per acre in two rows per 40-inch bed on Pullman Clay Loam at the High Plains Research Foundation in Plainview. Prior to planting, 100 pounds of 0-46-0 per acre were uniformly applied. The estimated level of preplant soil nitrogen was 12 pounds per acre.

The irrigation scheduling program is summarized in Table 9.41. The basic principle was to irrigate at two-, three-, or four-week intervals. Rainfall was a factor in this experiment. About 16.6 inches were recorded during the growing season. The estimated available water-holding capacity of soil at this site is 8 acre-inches in the upper 4 feet. Five levels of nitrogen were incorporated in the experimental design. Since each possible combination of the five irrigation and nitrogen treatments is included in the experiment, the experimental design is termed a randomized complete block with factorials. Plots were harvested November 20.

Treatment levels and corresponding root and sucrose yields are given in Table 9.42. The average yield response to nitrogen is strong for only irrigation regimes I_1 and I_4. For the other irrigation treatments, the yield response is weak and (or) inconclusive.

The average yield response to water is not only a function of total water applied but also the distribution of applications. Two comparisons of the relationships implicit in Tables 9.41 and 9.42 are of interest. The average yields associated with I_5 exceed those generated by I_1 at all levels of fertilization other than $N_5 = 200$ pounds per acre. The only difference in the irrigation programs is that with I_1 an additional 3.9 acre-inches were applied September 13. Based on these experimental data, this relatively late irrigation had a generally depressing effect on subsequent root yields. The other comparison is between yields corresponding to I_1 and I_2 where the latter involved about 5 acre-inches less water than I_1. Average yields generated by I_2, however, are higher at all nitrogen levels other than $N_5 = 200$ pounds per acre.

The quadratic function in (9.45) reflects the yield-water-nitrogen relationships for this experiment. The yield response to nitrogen does not exhibit any strong, consistent pattern. Predicted yields derived from (9.45) are in Table 9.43.

$$\text{yield} = -82.8235^{***} + 5.0909^{***}W_1 - .0507N_1 - .0678^{***}W_1^2$$
$$\text{(roots)} \quad (20.12) \quad (1.068) \quad (.041) \quad (.014)$$

$$- .00007N_1^2 + .0022^{**}W_1N_1 \tag{9.45}$$
$$(.0001) \quad (.001)$$

$$R^2 = .405 \quad F = 12.67^{***} \quad LOFF = 1.218$$

A P P E N D I X

Canonical Correlations and Joint Products

In collaboration with S. Roy Chowdhury and V. Nagadevara

In the body of this chapter, production functions were estimated for sugar beet tops and roots separately. These two commodities are joint products of sugar beet production. Hence, they also can be estimated jointly. By definition, a joint production function has two or more dependent variables in a single equation. Hence, ordinary least squares (OLS) may not be most appropriate as a predicting device when we are interested in both products. As shown by Vinod, the estimation of a joint production function by OLS can produce inconsistent and incorrect estimates of the parameters.[1] A preferred method, perhaps, is canonical correlation analysis as proposed by Hotelling.[2] Vinod and Kaminsky have adapted Hotelling's method in their analysis of joint production functions.[3] Vinod used all canonical correlations for estimating parameters of the production function. Since a joint production function is a single equation, a more logical method considers only the first canonical correlation. On the following pages, three forms of functions (quadratic, transcendental, and constant elasticity of substitution-transformation [CEST]) are tested. An improvement or variation from Kaminsky in the CEST form is made by using unequal (rather than equal) elasticities of transformation and substitution.

A joint production function can be described by the single equation

$$F(y_1, \ldots, y_p, x_1, \ldots, x_q) = 0 \tag{9.46}$$

where $y_1, \ldots, y_p$ are outputs and $x_1, \ldots, x_q$ are inputs. When F is linear, the statistical model for (9.46) is

$$\Sigma \alpha_i y_i = \Sigma \beta_i x_i + C + \epsilon \tag{9.47}$$

where C is a constant and ϵ is the disturbance term. Variables y_i and x_i can be represented in the matrix form

$$Y = \begin{bmatrix} y_{11} & y_{12} & \cdots & y_{1N} \\ y_{21} & y_{22} & \cdots & y_{2N} \\ \vdots & & & \\ y_{p1} & & \cdots & y_{pN} \end{bmatrix} \tag{9.48}$$

$$X = \begin{bmatrix} x_{11} & x_{12} & \cdots & x_{1N} \\ x_{21} & x_{22} & \cdots & x_{2N} \\ \vdots & & & \\ x_{q1} & & \cdots & x_{qN} \end{bmatrix} \tag{9.49}$$

where N is the number of observations. The sample observations in the matrices are assumed to be standardized by dividing deviations from sample means by sample standard deviations. Estimation of (9.47) by OLS is the same whether or not the outputs are from a joint production process. Also, by arbitrarily denoting one output variable as the dependent one and treating other variables as independent, OLS procedure can produce inconsistent and structurally incorrect estimates, as noted previously.

In canonical correlation, we form two canonical variables, u_1 and v_1, or the linear combinations of the standardized variables $(y_i - \bar{y}_i)/s_{y_i}$ and $(x_j - \bar{x}_j)/s_{x_j}$, where $\bar{y}_i, \bar{x}_j$ are the sample means and s_{y_i} and s_{x_j} are the sample standard deviations. Let y and x represent the vectors of the standardized variables. Then

$$u_1 = \theta'y, \text{ and} \tag{9.50}$$

$$v_1 = \delta'x \tag{9.51}$$

The parameters θ and δ are chosen to maximize the sample correlation between u_1 and v_1. If r_1 is that maximum correlation that is also known as the first canonical correlation, then

$$u_1 = r_1 v_1, \text{ and} \tag{9.52}$$

$$\theta'y = r_1 \delta'x \tag{9.53}$$

The estimates of α_i, β_i, and C in (9.47) are obtained by transforming the standardized variables back to the original ones:

$$\theta_1\left(\frac{y_1 - \bar{y}_1}{s_{y_1}}\right) + \theta_2\left(\frac{y_2 - \bar{y}_2}{s_{y_2}}\right) + \cdots + \theta_p\left(\frac{y_p - \bar{y}_p}{s_{y_p}}\right) = \delta_1\left(\frac{x_1 - \bar{x}_1}{s_{x_1}}\right)$$
$$+ \delta_2\left(\frac{x_2 - \bar{x}_2}{s_{x_2}}\right) + \cdots + \delta_q\left(\frac{x_q - \bar{x}_q}{s_{x_q}}\right) \tag{9.54}$$

The estimates of α, β, and C are:

$$\hat{\alpha}_i = \frac{\theta_i}{s_{y_i}}, \hat{\beta}_i = \frac{\delta_i}{s_{x_i}} \text{ and } \hat{C} = \sum \frac{\theta_i y_i}{s_{y_i}} - \sum \frac{\delta_i \bar{x}_i}{s_{x_i}} \tag{9.55}$$

The validity of the canonical correlation method lies in forming composite output and input variables and thus retaining the jointly dependent nature of the production process.

The mathematical derivations of obtaining θ, δ, and r_1 are given comprehensively in Anderson[4] and Press[5] and involve finding the characteristic roots and vectors of a certain matrix. It is also known that when there is only one output variable, the dependent one, OLS and canonical correlation yield the same results. Otherwise, canonical correlation and OLS are likely to give different results.

PRODUCTION FUNCTIONS

A brief description of the forms and characteristics of the production functions estimated follows.

Mundlak's[6] transcendental production function can be stated in the following implicit form:

$$G^g e^{g'G} \dots F^f e^{f'F} = A L^1 e^{1'L} \dots T^t e^{t'T} \tag{9.56}$$

where $G, \dots, F$ are output variables; $L, \dots, T$ are input variables; and g, $g', \dots, f, f', A, 1, 1', \dots, t, t'$ are parameters.

The relationship (9.56) is symmetric in inputs and outputs and is relatively simple. In logarithmic form it can be written as

$$g \log G + g'G + \cdots + f \log F + f'F = \log A + 1 \log L + 1'L + \cdots + t \log T + t'T \tag{9.57}$$

The relation (9.57) can be estimated by canonical correlation by considering $\log G, G, \dots, \log F, F, \log L, L, \dots, \log T, T$ as the variables. Fulfillment of the first-order and second-order conditions of profit maximization places certain restrictions on the parameters given in (9.58) through (9.61). Details as to the derivation of (9.58) through (9.61) are found in Mundlak.

$$g, \dots, f < 0 \tag{9.58}$$

$$1, \dots, t > 0 \tag{9.59}$$

$$g' > \left| \frac{g}{G} \right|, \dots, f' > \left| \frac{f}{F} \right| \tag{9.60}$$

$$1' > -\frac{1}{L}, \dots, t' > -\frac{t}{T} \tag{9.61}$$

When $g, \dots, f, 1, \dots, t$ are zero, the production function (9.56) exhibits Cobb-Douglas form both on the output and input side. This generalized

Cobb-Douglas form was used by Klein[7] who pointed out that the conventional Cobb-Douglas form on the output side alone renders a transformation curve with the wrong curvature and leads to minimum rather than maximum profits. Empirically, the transcendental form needs to be studied more for its suitability.

We also use the constant elasticity of substitution and transformation or CEST form of production function. Powell and Gruen[8] derived the family of constant elasticity of transformation (CET) production surfaces, which are of the form

$$g\, G^{1-k} + \cdots + f\, F^{1-k} = K \tag{9.62}$$

where $G, \ldots, F$ are output variables; $g, \ldots, f$ are parameters; and K is an index of a given resource bundle.

The main feature of (9.62) is its constant elasticity of transformation between any pair of outputs, given by $1/k$. The formulations of partial elasticity of transformation and elasticity of substitution given by Allen[9] are used throughout this appendix. By considering a similar form on the input side, Kaminsky[10] extended the CET form (9.62) to CEST form as

$$g\, G^{1-k} + \cdots + f\, F^{1-k} = C + 1\, L^{1-j} + \cdots + t\, T^{1-j} \tag{9.63}$$

where $L, \ldots, T$ are the input variables; $1, \ldots, t$ are the parameters; C is a constant; $\tau = 1/k$, τ being the constant elasticity of transformation; and $\sigma = 1/j$, σ being the constant elasticity of substitution.

For symmetry, he considered both the elasticity of transformation and the elasticity of substitution to be the same, which means

$$\tau = -\sigma \text{ or} \tag{9.64}$$

$$k = -j \tag{9.65}$$

Constant elasticities of transformation and substitution are considered below, but they do not necessarily have the same values. In other words

$$\tau = -\sigma \text{ or } \tau \neq -\sigma \tag{9.66}$$

The profit maximization condition for (9.63) requires k to be negative, j to be positive, and $g, \ldots, f > 0$.

Estimation of (9.63) by canonical correlation requires assigning values for k and j and treating $G^{1-k}, \ldots, F^{1-k}, L^{1-j}, \ldots, T^{1-j}$ as variables. Different combinations of k and j are tried and the one with the highest canonical correlation is finally chosen. This procedure is pragmatic for the situation.

The form (9.63) is suitable for representing usual transformation and substitution curves. When τ and σ are infinite, the production function becomes linear transformation and substitution curves. For other values of τ and σ we have convex and concave transformation and substitution curves, the curvatures being determined by the particular values of τ and σ chosen.

The third type of production function considered is quadratic. Specifically, the following form of quadratic function is considered

$$\alpha_1 y_1^2 + \alpha_2 y_2^2 + \alpha_3 y_1 + \alpha_4 y_2 + \alpha_5 y_1 y_2 = C + \beta_1 x_1^2 + \beta_2 x_2^2$$

$$+ \beta_3 x_1 + \beta_4 x_2 + \beta_5 x_1 x_2 \quad (9.67)$$

where y_1, y_2 are output variables; x_1, x_2 are input variables; and $\alpha_1, \alpha_2, \alpha_3$, $\alpha_4, \alpha_5, C, \beta_1, \beta_2, \beta_3, \beta_4, \beta_5$ are parameters.

In (9.67) we allowed interaction coefficients to occur both on the output and input side. The production relation (9.67) can be stated in the implicit form as

$$F = \alpha_1 y_1^2 + \alpha_2 y_2^2 + \alpha_3 y_1 + \alpha_4 y_2 + \alpha_5 y_1 y_2 - C - \beta_1 x_1^2 - \beta_2 x_2^2$$

$$- \beta_3 x_1 - \beta_4 x_2 - \beta_5 x_1 x_2 = 0 \quad (9.68)$$

The Lagrangian function for profit maximization is given as

$$H = p_1 y_1 + p_2 y_2 - q_1 x_1 - q_2 x_2 + \lambda F \quad (9.69)$$

where p_1, p_2, and $q_1 q_2$ are the prices of outputs and inputs and λ is the vector of Lagrangian multipliers. The first-order conditions for profit maximization are

$$\frac{\partial H}{\partial y_1} = p_1 + \lambda F y_1 = 0 \quad (9.70)$$

$$\frac{\partial H}{\partial y_2} = p_2 + \lambda F y_2 = 0 \quad (9.71)$$

$$\frac{\partial H}{\partial x_1} = -q_1 + \lambda F x_1 = 0$$

$$\frac{\partial H}{\partial x_2} = -q_2 + \lambda F x_2 = 0 \quad (9.72)$$

$$\frac{\partial H}{\partial \lambda} = F = 0$$

In (9.72) $F y_i$ and $F x_i$ are partial derivatives of F with respect to y_i and x_i. The second-order conditions for profit maximization require that the relevant bordered Hessian determinants alternate in sign:

$$\begin{vmatrix} \lambda F y_1 y_1 & \lambda F y_1 y_2 & F y_1 \\ \lambda F y_2 y_1 & \lambda F y_2 y_2 & F y_2 \\ F y_1 & F y_2 & 0 \end{vmatrix} > 0 \quad \begin{vmatrix} \lambda F y_1 y_1 & \lambda F y_1 y_2 & \lambda F y_1 x_1 & F y_1 \\ \lambda F y_2 y_1 & \lambda F y_2 y_2 & \lambda F y_2 x_1 & F y_2 \\ \lambda F x_1 y_1 & \lambda F x_1 y_2 & \lambda F x_1 x_1 & F x_1 \\ F y_1 & F y_2 & F x_1 & 0 \end{vmatrix} < 0$$

$$(9.73)$$

and

$$\begin{vmatrix} \lambda Fy_1y_1 & \lambda Fy_1y_2 & \lambda Fy_1x_1 & \lambda Fy_1x_2 & Fy_1 \\ \lambda Fy_2y_1 & \lambda Fy_2y_2 & \lambda Fy_2x_1 & \lambda Fy_2x_2 & Fy_2 \\ \lambda Fx_1y_1 & \lambda Fx_1y_2 & \lambda Fx_1x_1 & \lambda Fx_1x_2 & Fx_1 \\ \lambda Fx_2y_1 & \lambda Fx_2y_2 & \lambda Fx_2x_1 & \lambda Fx_2x_2 & Fx_2 \\ Fy_1 & Fy_2 & Fx_1 & Fx_2 & 0 \end{vmatrix} > 0 \qquad (9.74)$$

where Fy_iy_j, Fy_ix_j, and Fx_ix_j are second partial derivatives of F.

$$\begin{aligned} Fy_1 &= 2\alpha_1 y_1 + \alpha_3 + \alpha_5 y_2 & Fy_1y_1 &= 2\alpha_1 \\ Fy_2 &= 2\alpha_2 y_2 + \alpha_4 + \alpha_5 y_1 & Fy_1y_2 &= \alpha_5 \\ Fx_1 &= -2\beta_1 x_1 - \beta_3 - \beta_5 x_2 & Fy_1x_1 &= 0 \\ Fx_2 &= -2\beta_2 x_2 - \beta_4 - \beta_5 x_1 & Fy_1x_2 &= 0 \\ Fy_2y_1 &= \alpha_5 & Fx_1y_1 = 0 \quad & Fx_2y_1 = 0 \\ Fy_2y_2 &= 2\alpha_2 & Fx_1y_2 = 0 \quad & Fx_2y_2 = 0 \\ Fy_2x_1 &= 0 & Fx_1x_1 = -2\beta_1 \quad & Fx_2x_1 = -\beta_5 \\ Fy_2x_2 &= 0 & Fx_1x_2 = -\beta_5 \quad & Fx_2x_2 = -2\beta_2 \end{aligned}$$

(9.75)

It is necessary for a profit-maximizing point with nonnegative outputs and inputs to satisfy (9.72) and (9.73). The second-order conditions (9.73) can be ensured if the transformation curves between pairs of outputs and substitution curves between pairs of inputs are well-behaved, that is, they have the usual convexity and concavity. Production functions of the form (9.71) can be easily estimated by canonical correlation.

Empirical results

The basic data relates to two output variables (roots and tops of sugar beets) and two input variables (water and nitrogen). The data are obtained from the Yuma-Mesa 1969–70 experiments. The units of measurement of variables are as follows: R = sugar beet roots (tons per acre); T = sugar beet tops (tons per acre); N = nitrogen applied (pounds per acre); and W = water applied (acre-inches per acre).

The estimate of the transcendental form (9.57) with R, T as outputs and N, W as inputs by canonical correlation is

$$\log R - 0.0386\,R + 0.4238\,\log T + 0.0245\,T = 52.5796 - 17.1922\,\log W$$

$$+ 0.3107\,W + 0.4396\,\log N - 0.0007\,N \qquad (9.76)$$

where the canonical correlation = 0.9148.

The profit maximization conditions (9.58) and (9.59) for the transcendental form are violated in relation (9.76) as the coefficients attached to log

R and log T are not negative and the coefficient attached to log W is not positive. This particular form, therefore, is unsuitable for economic analysis.

As stated in (9.63) the CEST production function is of the form

$$r R^{1-k} + t \, T^{1-k} = C + w \, W^{1-j} + n \, N^{1-j} \qquad (9.77)$$

where r, t, C, w, and n are parameters and k and j are the reciprocals of the elasticities of transformation and substitution, respectively. To choose an appropriate estimate, canonical correlation is applied on (9.77) with different combinations of elasticity of transformation and elasticity of substitution. Table 9.44 summarizes the results and shows that for a particular size of elasticity of transformation (elasticity of substitution), as the size of elasticity of substitution (elasticity of transformation) increases, the overall goodness of fit, as measured by canonical correlation, also increases. Since the improvement is very little when the elasticity of transformation-substitution is more than five and since a linear transformation-substitution curve corresponding to size ∞ is to be avoided, an elasticity of five is finally chosen. The canonical correlation estimate corresponding to an elasticity of five is

$$R^{6/5} + 1.6885T^{6/5} = -292.4561 + 11.8735W^{4/5} + 0.1518N^{4/5}$$
$$\qquad\qquad\qquad\qquad (6.9782) \qquad\quad (1.6807) \qquad\qquad (9.78)$$

where the canonical correlation = 0.8815 and R^2 = .7770.

Finally, the estimate of the quadratic production function is

$$R^2 + 1.5122T^2 = -1418.9372 + 0.4383W^2 + 0.0014N^2 - 0.0041WN$$
$$\qquad\qquad\qquad\qquad (3.4920) \qquad (.3434) \qquad (-.1257)$$
$$(9.79)$$

where the canonical correlation = .8443 and R^2 = .7128.

The canonical correlation method does not provide t values (for testing significance of the estimated coefficients) and multiple correlation. The procedure followed here and given by Kaminsky[11] is to collapse the variables on the left-hand side of the relations (9.78) and (9.79) as one variable and regress on the variables on the right-hand side. This provides t values, which are given in parentheses in (9.78) and (9.79). The above procedure is based on the fact that canonical correlation is equivalent to a multiple regression on one side of the production surface once the composite variate on the other side is determined.

APPLICATIONS

Economic analyses considering the marginal products, the marginal rate of substitution, the marginal rate of transformation, and profit maximization equations are summarized in this section.

For particular values of $T = T_0$ and $R = R_0$, Equation (9.79) will take the following forms:

$$R^2 = (-1418.9372 - 1.5122T_0^2) + 0.4383W^2 + 0.0014N^2 - 0.0041WN \quad \textbf{(9.80)}$$

$$1.5122T^2 = (-1418.9372 - R_0^2) + 0.4383W^2 + 0.0014N^2 - 0.0041WN \quad \textbf{(9.81)}$$

The marginal products and the marginal rate of substitution are given by

$$\frac{\partial R}{\partial W} = \frac{0.8766W - 0.0041N}{2R} \qquad \textbf{(9.82)}$$

$$\frac{\partial R}{\partial N} = \frac{0.0028N - 0.0041W}{2R} \qquad \textbf{(9.83)}$$

$$\frac{\partial T}{\partial W} = \frac{0.8766W - 0.0041N}{3.0244T} \qquad \textbf{(9.84)}$$

$$\frac{\partial T}{\partial N} = \frac{0.0028N - 0.0041W}{3.0244T} \qquad \textbf{(9.85)}$$

$$MRS = -\frac{\partial W}{\partial N} = \frac{0.0028N - 0.0041W}{0.8766W - 0.0041N} \qquad \textbf{(9.86)}$$

Similarly, the marginal rate of transformation is given by

$$MRT = -\frac{dT}{dR} = \frac{R}{1.5122T} \qquad \textbf{(9.87)}$$

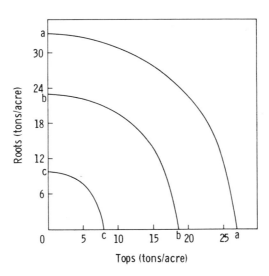

FIG. 9.8. Product transformation curve of CEST production function (Yuma Mesa, Arizona, 1969–70).

The first-order conditions of profit maximization are given in

$$MRT = \frac{R}{1.5122T} = \frac{P_1}{P_2}$$
(9.88)

$$p_1 \left(\frac{0.8766W - 0.0041N}{2R} \right) = r_1$$
(9.89)

$$p_1 \left(\frac{0.0028N - 0.0041W}{2R} \right) = r_2$$
(9.90)

$$p_2 \left(\frac{0.8766W - 0.0041N}{3.0244T} \right) = r_1$$
(9.91)

$$p_2 \left(\frac{0.0028N - 0.0041W}{3.0244T} \right) = r_2$$
(9.92)

$$MRS = -\frac{\partial W}{\partial N} = \frac{0.0028N - 0.0041W}{0.8766W - 0.0041N} = \frac{r_2}{r_1}$$
(9.93)

where p_1, p_2, r_1, and r_2 are prices of R, T, W, and N. Given the prices of outputs and inputs, if there is a nonnegative output-input combination that satisfies both the first- and second-order conditions, then that point must be a profit-maximizing point. With respect to the CEST estimate (9.78) we thus have

$$\frac{\partial R}{\partial W} = \frac{23.7470W^{-1/5}}{3R^{1/5}}$$
(9.94)

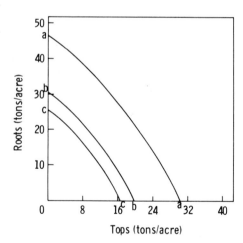

FIG. 9.9. Product transformation curve of quadratic production function (Yuma Mesa, Arizona, 1969-70).

$$\frac{\partial R}{\partial N} = \frac{0.3036N^{-1/5}}{3R^{1/5}} \tag{9.95}$$

$$\frac{\partial T}{\partial W} = \frac{4.6880W^{-1/5}}{T^{1/5}} \tag{9.96}$$

$$\frac{\partial T}{\partial N} = \frac{.0599N^{-1/5}}{T^{1/5}} \tag{9.97}$$

$$MRS = -\frac{\partial W}{\partial N} = \frac{.0128N^{-1/5}}{W^{-1/5}} \tag{9.98}$$

$$MRT = -\frac{dT}{dR} = \frac{R^{1/5}}{1.6885^{1/5}} \tag{9.99}$$

The first-order profit-maximizing conditions for (9.78) can be easily derived for given prices of outputs and inputs. Product transformation curves for (9.78) and (9.79) for different levels of input combinations are given in Figures 9.8 and 9.9, respectively. In both figures, cc is for $W = 59.0$ and $N = 86.0$, bb is for $W = 66.48$ and $N = 236.0$ (mean level), and aa is for $W = 75.2$ and $N = 386.0$. The well-behaved shapes of the transformation curves (Figures 9.8 and 9.9) ensure the second-order conditions of profit maximization.

GENERALIZED PRODUCTION FUNCTIONS

Production functions quantifying yield-water-nitrogen relationships in various degrees were presented in preceding chapters. These estimated relationships are to some unknown degree site specific and, therefore, not necessarily applicable to other locations. As noted in Chapter 3, when the water treatments are defined in terms of soil moisture conditions, some transferability of results may be expected. This potential transferability occurs because soil moisture tension levels are affected by soil characteristics and climatic conditions. Other indigenous factors such as seed variety, plant population, levels of preplant soil fertility and managerial inputs must also be taken into account.

If the experimental data at each site could be aggregated and synthesized into "generalized" input-output relationships for a limited number of variables, the potential for estimating yields at other nonexperimental sites may be augmented. To this end, experiments were conducted at sites having varying soil and climatic conditions. Corn experiments, for example, were set up in five states and at a total of nine sites within the five states. Of the four sites in Arizona, the Yuma Mesa and Yuma Valley sites are about ten miles apart; however, the soil texture differs considerably between the two sites. The soil type at Yuma Mesa is Superstition Fine Sand with the sand content between 70 and 80 percent and an available water-holding capacity of only 3.1 inches in the upper 4 feet. In comparison, the soil type at the Yuma Valley site is Glendale Silty Clay Loam having a sand content of less than 10 percent and an available water-holding capacity of 9.9 inches in the upper 4 feet. As a consequence, heavier and more frequent applications of water and fertilizer were required at the Yuma Mesa site. Soil nutrients are easily leached from this sandy soil. For further contrast, the Pima Clay Loam variant and the groundwater at the Safford site have a relatively high salt content, particularly sodium. The available water-holding capacity of this soil is 7.5 inches in the top 4 feet.

The geographical representation of corn experiments provides data for a relatively wide range of climatic conditions. Climatic conditions in California and Arizona are considerably different from those in Colorado, Kansas, and Texas. Precipitation levels, length of growing season, and temperature and humidity conditions are important determinants of yield.

These soils and climatological data provide a basis for developing generalized production functions. These functions for individual crops are generalized in the sense that coefficients are estimated for the effect of water, nitrogen, soil, and weather factors in the aggregated experiments. With inclusion of the site variables, the generalized functions are potentially useful for predicting yields at other geographical locations. The degree of transferability, however, is limited. Since the experiments were conducted at locations in western and southwestern United States, the generalized functions are more relevant for predicting yields in, for example, New Mexico than in Illinois.

SPECIFICATION OF SITE VARIABLES

Some of the site variables are quantifiable; others are not. In aggregating experimental data from individual sites, managerial and labor inputs are expected to be different. Plant varieties differ as do chemical compositions of the fertilizer treatments. Reasonably comparable data on several soil characteristics and weather factors are available for all sites. These are qualified as being reasonably comparable because the soil analyses were conducted in different laboratories, and while analytical procedures tend to be comparable, they are not necessarily strictly identical.

Soil characteristics

Since the soil analyses for individual experimental sites were not fully standardized, reasonably comparable data are available only for the following 12 soil characteristics; percentage sand; percentage silt; percentage clay; pH; electrical conductivity (EC); cation exchange capacity (CEC); available water-holding capacity (AWC); hydraulic conductivity (HC); and measurements of K (potassium), Ca (calcium), Mg (magnesium), and Na (sodium) in the soil saturation extract. Topographical features were not included. The implicit assumption is made that topography is uniform and no drainage problems exist.

Preliminary analyses indicated that correlations between yields and soil characteristics were higher when data for soil characteristics at various soil depths were aggregated into one weighted value for each characteristic rather than using individual measurements throughout the soil profile. Each agronomist conducting experiments was requested to provide estimates of weights reflecting the relative importance of the specified soil layer on plant growth and yield. In most cases these weights were based on the relative distribution of plant roots through the effective root zone. To clarify this approach, the estimated percent sand at four soil levels for the 1971 wheat and 1971–72 sugar beet experiments at Mesa, Arizona, is given in Table 10.1. As is apparent, the top foot of soil was given a slightly higher weight for the sugar beet experiment compared with 4 for wheat. The weighted average of percent sand applicable to the wheat experiment was derived as 56.8 percent compared with 53.96 percent for the sugar beet experiment. The weights estimated by the agronomists vary with the experimental site and the crop grown. These are listed in Appendix Table 1. The weighted values for the most important soil characteristics corresponding to each experiment are summarized in Appendix Table 2.

In addition to deriving weighted estimates of selected soil characteristics, a few of the soil measurements were further classified according to soil-engineering classes developed by the Soil Conservation Service, U.S. Department of Agriculture.[1] These classifications may be helpful to agronomists and soil scientists preferring a range of measurements for soil characteristics rather than point estimates. Measurements are always subject to some variability and the use of ranges or classes permits allowance for this variability. The classes of measurements and the coded values assigned to each class are given in Table 10.2. The neutral or normal range is assigned a coded value equal to zero.

Weather factors

To reflect climatic factors at individual experimental sites, pan evaporation data were used as a proxy embodying temperature, wind, and humidity conditions during the growing season. Daily pan evaporation measurements are available for most sites. The impact of climatic factors on crop growth and yields is not equally distributed throughout the growing season. For example, climatic factors during pollination and silking for corn should be weighted more heavily than those close to harvest. Agronomists conducting the experiments were requested to specify the "critical periods" during which climatic factors most significantly affect final yields. There was general agreement that the most critical period for corn was during silking. For wheat, the "boot-dough" stage was mentioned most frequently. Based on a synthesis of the agronomists' estimates of the "critical period" for cotton, the crucial period is approximately a 20-day period during the flowering stage beginning about 105 days prior to the first harvest. Because of the variety of comments for sugar beet production, no particular "critical period" was discernible. Planting and harvesting dates and estimates of these "critical periods" are given in Appendix Table 3.

In the Texas, Kansas, and Colorado experiments, precipitation levels were sufficiently high so that at least some precipitation should be added to the irrigation treatment levels. The question is one of how much. For example, 0.10 inch, other things equal, has less impact than 0.80 inch. Further, any precipitation occurring the day following an irrigation treatment likely has relatively little impact on plant growth. As a first approximation and rule of thumb, only that precipitation equal to or exceeding 0.25 inch at any occurrence was added to the irrigation treatment levels. Preplant irrigations raising the soil moisture level to field capacity were also included. Experimental plots were basined to minimize any runoff from water application and (or) precipitation.

ANALYTICAL PROCEDURE

Several sequential procedures were employed in constructing the generalized production functions. In addition to the previously discussed specification bias associated with estimating individual production functions, estimating generalized production functions also introduces the possibility for aggregation bias. In the functions developed here, aggregation bias arises from combining experimental data generated by different experimental designs, managerial inputs, and plant varieties. When a dummy variable approach is used, these differences are embodied and, in a sense, isolated within the dummy variables.

Alternative production functions were fitted to each set of experimental data at each site. The next step in the sequential estimation procedure was to examine the question of whether the derived regression coefficients, for example, the linear coefficients for water, represent samples from a multivariate normal distribution or samples from different populations. Cady and Fuller developed a procedure for examining this question.[2] This test to assess the variability of site responses was applied to corn and cotton. Corn experiments were conducted in five states and represent the maximum among-site variability. Cotton experiments were conducted only in California and Arizona and, therefore, should represent the minimum among-site variability. The Cady and Fuller test involves conducting a significance test (F test) for each vector of derived regression coefficients. If the derived F is nonsignificant, variation among, for example, the regression coefficients for water can be attributed to experimental error; and the site variables are of little importance in explaining variability among the water coefficients. If the derived F is statistically significant, site variables should be included when experimental data from the various sites are aggregated. Some of the derived F's were significant for both corn and cotton. The next question is: What site variables should be included?

Comparable data for nine site variables were available—eight soil variables plus the pan evaporation data summarizing the important climatological factors. Measurements of K, Ca, Mg, and Na in the soil saturation extract were in different units, and sufficient information was not available for transforming the individual measurements to a common base. Weighted values for the eight soil variables were derived according to the procedure outlined in Table 10.1. Two weather variables were defined. The first represented the summation of pan-evaporation data from planting up to about three weeks prior to harvest. This variable would also reflect the length of the growing season. The second weather variable was defined as the summation of pan-evaporation data during the "critical period" of crop growth, as defined by agronomists conducting the experiments.

A matrix of simple correlations among the eight soil variables was derived. Hydraulic conductivity or available water-holding capacity was highly correlated with site variables other than pH and electrical conductivity. Since hydraulic conductivity and available water-holding capacity tend to be highly correlated, only one was used in constructing generalized production functions.

Interactions among certain site variables, water, and nitrogen were hypothesized to be important. Voss and Pesek have suggested a procedure for determining the possibly important interactions.[3] In brief, the procedure involves estimating the partial correlation coefficients among each vector of regression coefficients, for example, W, and the site variables. If the partial correlation between water and available water-holding capacity is statistically significant at some predetermined level, an interaction term for these two variables is a possible candidate for inclusion in the generalized production function.

Estimation of the generalized functions is based on ordinary least-squares techniques. Alternative polynomial forms such as quadratic, square root, three-halves, and 1.5 polynomial were routinely estimated. Various estimations involving combinations of site variables and interaction terms were also made. Only one of the alternative polynomials is specified for each crop. This is the model that generates the best test statistics and is most consistent with hypothesized relationships.

CORN (GRAIN)

Using the above analytical procedure, the 1.25 polynomial in (10.1) was fitted to experimental data from nine sites. Data were aggregated from the 1968 Colorado; 1970 California; 1971 Kansas; 1970 Plainview, Texas; 1971 Plainview (Lake Site), Texas; and 1970 Mesa, 1972 Safford, 1970 Yuma Mesa, and 1970 Yuma Valley, Arizona, experiments. The variables in (10.1) are defined as

$$
\begin{aligned}
\text{yield} = 3789.3801 &\overset{**}{+} 4047.1592 \overset{***}{P} + 579.4986 \overset{***}{W_1} + 29.2230 \overset{***}{N} \\
(1627.88)\quad &(442.12)\qquad (85.66)\qquad (5.495) \\[6pt]
&\overset{***}{-}\, 262.8332 W_1^{1.25} - 5.8989 \overset{***}{N^{1.25}} - .0288 W_1 N \\
&(35.97)\qquad\quad (1.261)\qquad\ (.039) \\[6pt]
&\overset{***}{-}\, 4658.1284 pH + 1961.5155 \overset{***}{EC} - 149.4873 \overset{**}{AWC} \\
&(649.64)\qquad\quad (171.01)\qquad (73.57) \\[6pt]
&\overset{***}{-}\, 2596.8452 PE2 + 45.0936 \overset{***}{W_1 PE2} \\
&(412.56)\qquad\quad (8.532)
\end{aligned}
\tag{10.1}
$$

$$R = .652 \quad \overset{***}{F} = 87.84$$

where

  yield = pounds per acre of corn (grain) at 15.5 percent moisture;
  P = plant population, in 10,000 per acre;
  W_1 = inches per acre water applied plus total rainfall exceeding 0.25 inches at any occurrence;
  N = pounds per acre of nitrogen applied;
  $pH = (pH - 7.0)^2$;
  EC = electrical conductivity in millimhos per centimeter;
  AWC = available water-holding capacity, in acre-inches in upper 4 feet; and
  $PE2$ = sum of pan-evaporation data over 10-day silking period, defined as the "critical period," in inches of water evaporated.

Except for the water × nitrogen interaction variable, all estimated coefficients are statistically significant at the 5 percent or higher level. The signs for most coefficients are consistent with hypothesized relations. A negative coefficient, however, would be expected for EC. That is, as the salinity content of the soil increases, yield would be expected to decline. A positive relationship between the level of available water-holding capacity (AWC) and yield would be hypothesized. The estimated coefficient in (10.1), however, is negative. Coefficients for the variables including the weather factor are consistent with expectations. As high temperatures and (or) low humidity during the "critical period" of pollination and silking cause $PE2$ to increase, yields would be adversely affected and a negative coefficient for $PE2$ would be expected. As $PE2$ increases, the water requirements of plants also increase. Consequently, the $W_1 \times PE2$ variable has a positive coefficient.

Based upon Equations (6.39) and (6.45) from Chapter 6, the experiments for the Safford, Arizona, and Plainview (Lake), Texas, sites were not particularly good. When these data were omitted, the 1.5 polynomial in (10.2) was fitted to the data for the remaining seven sites. A modified version of (10.2) including the $W_1 \times PE2$ interaction term was also estimated; however, the derived coefficient for the interaction variable was not statistically significant at conventional levels. In comparing (10.2) with (10.1), the coefficient for the pH variable now has a positive sign; however, the estimated coefficient for AWC has changed from a negative to a positive value. Almost all other properties are similar. The statistical significance of the plant population coefficient in (10.2) has declined relative to (10.1).

$$
\begin{aligned}
\text{yield} = &\ 10473.5073 \overset{***}{} + 1081.9211\overset{*}{P} + 716.5545\overset{***}{W_1} + 19.1775\overset{***}{N} \\
&\ (1432.13) \quad (577.21) \quad\quad (51.03) \quad\quad (3.039) \\[8pt]
&\ - 80.4347\overset{***}{W_1^{1.5}} - .8706\overset{***}{N^{1.5}} + .0385W_1N + 6869.7026\overset{***}{pH} \\
&\ (5.902) \quad\quad (.161) \quad\quad (.050) \quad\quad (1087.90) \\[8pt]
&\ + 488.3393\overset{**}{EC} + 607.8084\overset{***}{AWC} - 5465.4571\overset{***}{PE2} \\
&\ (199.94) \quad\quad (89.71) \quad\quad (364.70)
\end{aligned}
\qquad \textbf{(10.2)}
$$

$$R^2 = .718 \quad F = \overset{***}{105.24}$$

The across-site variability can also be represented by a system of dummy variables. Let the system be constructed as

	X_1	X_2	X_3	X_4	X_5	X_6
1968 Colorado	0	0	0	0	0	0
1970 California	1	0	0	0	0	0
1971 Kansas	0	1	0	0	0	0
1970 Texas	0	0	1	0	0	0
1970 Mesa, Arizona	0	0	0	1	0	0

$$X_1 \quad X_2 \quad X_3 \quad X_4 \quad X_5 \quad X_6$$

	X_1	X_2	X_3	X_4	X_5	X_6
1970 Yuma Mesa, Arizona	0	0	0	0	1	0
1970 Yuma Valley, Arizona	0	0	0	0	0	1

where the 1968 Colorado experiment is arbitrarily chosen as the base or reference site. Since each of the estimated coefficients for the dummy variables in (10.3) is statistically significant at the 1 percent or higher level, the site effect at each of the remaining sites is significantly different from the site effect at Colorado. As noted earlier, the dummy variables would include the effect of differences in experimental design, soil and weather factors, seed varieties, and managerial levels. But the interest is in isolating, where possible, specific site effects such as soil and weather factors. In this context, (10.2) provides considerably more information than (10.3).

$$\text{yield} = 1515.5597^{***} - 1487.1796^{***}X_1 - 3772.2142^{***}X_2$$
$$\quad\quad (361.32) \quad\quad (252.66) \quad\quad\quad (280.62)$$

$$- 10622.4597^{***}X_3 - 7878.3318^{***}X_4 - 3751.7872^{***}X_5$$
$$\quad (409.80) \quad\quad\quad (315.13) \quad\quad\quad (286.50)$$

$$- 6594.0042^{***}X_6 + 970.7438^{***}W_1 + 30.1923^{***}N \quad\quad\quad\quad (10.3)$$
$$\quad (267.62) \quad\quad\quad (83.45) \quad\quad\quad (4.544)$$

$$- 289.0799^{***}W_1^{1.25} - 6.2153^{***}N^{1.25} + .0261W_1N$$
$$\quad (29.30) \quad\quad\quad (1.075) \quad\quad\quad (.043)$$

$$R^2 = .796 \quad\quad F = 146.29^{***}$$

Finally, one additional variation was estimated where the site variables were coded according to the classes specified in Table 10.2. The quantified version of this model is in (10.4). The subscript c denotes that the variable has been included in the analysis at its coded value.

In comparing (10.2), (10.3), and (10.4), it appears that (10.2) provides the best results and the most information. Predicted yields derived from (10.2) are listed in Table 10.3. The average and range of observed yields for experiments reflect I_3N_1, I_3N_3, and I_3N_5 treatment combinations. The corresponding predicted yields and the 95 percent confidence limits for each predicted yield are listed by treatment combination. One method for attempting to validate a model is to use it to test data not incorporated within the model. The points of interest are a comparison of the observed average yield with the point estimate of predicted yield and whether the observed average yield falls within the 95 percent confidence limits for the predicted yields. If the tolerance level were increased from 5 to 10 percent, the confidence interval for predicted yields would become wider. In this validation, we use equation (10.2) to predict yield outcomes at various locations, then compare the predictions with actual yield outcome.

$$\text{yield} = 7048.1683^{***} + 3068.9985^{***}P + 371.2464^{***}W_1$$
$$(1248.11) \quad (536.57) \quad (52.18)$$

$$+ 18.5264^{***}N - 40.2879^{***}W_1^{1.5} - .9674^{***}N^{1.5}$$
$$(3.143) \quad (6.661) \quad (.166)$$

$$+ .1099^{**}W_1N + 80.0911pH_c + 4301.3726^{***}EC_c \quad \textbf{(10.4)}$$
$$(.052) \quad (483.74) \quad (389.84)$$

$$- 33.2797AWC_c - 3098.8849^{***}PE2$$
$$(167.52) \quad (148.03)$$

$$R^2 = .698 \quad F = 95.68^{***}$$

As one step in such a validation, observed yields for the Colorado experiment are averaged across six observations per treatment combination. Average observed yields are relatively close to predicted yields; they also fall within the 95 percent confidence interval for the respective predicted yields. Because of the experimental designs used in the California and Kansas experiments, only one or in some cases two observations per treatment combination are available. Consequently, comparisons between observed and predicted yields are less meaningful for these experiments.

For the 1970 experiment at Mesa, Arizona, predicted yields consistently overestimate observed yields. Recall that abnormally high temperatures and low humidity during pollination raised water requirements but also reduced yields. The relatively high predicted yields appear to result from the high levels of water applied.

Observed and predicted yields are reasonably similar for the Yuma Mesa Valley tests. For these experiments, observed yields are based upon six observations per treatment combination. The largest discrepancy occurs for $I_3 = 30.5$ and $N_1 = 0$ treatment combination at Yuma Valley; however, this difference is only 424 pounds or 7.6 bushels per acre. Average observed yields for the other treatment combinations fall within the 95 percent confidence interval for their respective predicted yields.

Predicted yields for the 1970 Texas experiments at other than the highest level of fertilization are consistently below the average observed yields. These averages are based on four observations per treatment combination. There is no apparent reason for these differences.

CORN (SILAGE)

The generalized function of corn (silage) in (10.5) is based on data from the 1968 Colorado test and the 1971 Mesa, 1970 Yuma Mesa, and 1970 Yuma Valley experiments in Arizona. Due to some interrelationships within the model, the $X'X$ matrix becomes singular when variables for both P and AWC are included. Consequently, the plant population variable has been omitted in (10.5). When the corn silage data are aggregated, a better fit is obtained if the weather variable is defined as $PE1$ rather than $PE2$. $PE1$ represents the sum of daily pan-evaporation data from planting up to 2-3 weeks prior to harvest, thus also reflecting the length of the growing season. Each of the estimated coefficients in (10.5) has the correct, that is, expected sign and is statistically significant at the 10 percent or higher level.

$$\text{yield} = 14558.9721^* - 3481.2905^{**}W_1 - 64.8970^{***}N$$
$$\phantom{\text{yield} =} (8309.70) \quad\quad (1558.88) \quad\quad (10.94)$$

$$+ 13192.7149^{***}W_i^{.5} + 518.0722^{**}N^{.5} + 244.7131^{***}W_i^{.5}N^{.5}$$
$$ (2162.70) \quad\quad\quad (251.78) \quad\quad\quad (54.71)$$

$$- 70254.4047 pH - 3386.9781^{***}EC + 3000.467^{***}AWC \quad\quad (10.5)$$
$$ (14701.52) \quad\quad (1098.27) \quad\quad (489.28)$$

$$+ 58.4337^{**}W_1 PE1$$
$$ (27.48)$$

$$R^2 = .783 \quad\quad F = 84.37^{***}$$

When the soil variables are coded, the square root function is reestimated as in (10.6). In comparison with (10.5), the standard errors for several estimated coefficients are relatively high so that their statistical significance is lower.

$$\text{yield} = -1568.4564 - 4839.6651^*W_1 - 64.1500^{***}N$$
$$\phantom{\text{yield} =} (6365.93) \quad\quad (2763.62) \quad\quad (10.97)$$

$$+ 17086.8203^{***}W_i^{.5} + 478.2496^*N^{.5} + 249.2383^{***}W_i^{.5}N^{.5}$$
$$ (6207.55) \quad\quad\quad (253.61) \quad\quad\quad (54.94)$$

$$- 50599.3972^{***}pH_c - 25179.4591^{***}EC_c \quad\quad\quad (10.6)$$
$$ (16924.23) \quad\quad (3325.77)$$

$$+ 796.9534 AWC_c + 76.3139^*W_1 PE1$$
$$ (2971.45) \quad\quad\quad (42.32)$$

$$R^2 = .783 \quad\quad F = 84.00^{***}$$

Using (10.5) as a predicting equation, predicted yields and observed yields for the $I_3 N_1$, $I_3 N_3$, and $I_3 N_5$ treatment combinations are listed in Table 10.4. Predicted and observed yields are relatively close except for the high nitrogen treatment at Mesa. Note that the range of observed yields for this treatment combination is relatively wide.

WHEAT

Data from five wheat experiments were aggregated in the estimation of (10.7). These experiments were the Colorado trial plus the 1970-71 Mesa, 1970-71 Safford, 1971-72 Yuma Mesa, and 1971-72 Yuma Valley experiments in Arizona. In fitting alternative models to the wheat data, better results were obtained when a variable for the hydraulic conductivity of the soils (HC) was used rather than available water-holding capacity (AWC).

Variable $PE2$ is defined as the sum of daily pan-evaporation data during the "boot-dough" stage. The estimated "critical period" covers a 55-day interval beginning 75 days prior to harvest. All estimated coefficients in (10.7) are statistically significant at the 5 percent or higher level.

$$\text{yield} = 2799.8713^{***} + 269.6430^{***}W_1 + 17.2385^{***}N - 11.6997^{**}W_1^{1.5}$$
$$\qquad (422.93) \qquad (41.51) \qquad (3.016) \qquad (5.467)$$

$$\qquad - .6902^{***}N^{1.5} - .0953^{**}W_1N - 1338.3341^{***}pH - 616.990^{***}EC$$
$$\qquad\quad (.176) \qquad\quad (.046) \qquad\quad (363.02) \qquad\quad (38.51) \qquad (10.7)$$

$$\qquad - 932.1562^{***}HC - 1.7171^{***}W_1PE2$$
$$\qquad\quad (70.48) \qquad\qquad (.549)$$

$$R^2 = .746 \qquad F = 84.40^{***}$$

The negative sign for the $W_1 \times N$ variable indicates that W_1 substitutes for N. That is, within the range of treatment levels represented by the five wheat experiments, use of additional fertilizer requires a smaller application of water. With the exception of the $W_1 \times PE2$ variable, the signs for the derived coefficients are consistent with hypothesized values.

When the model is estimated in terms of coded site variables, (10.7) is reestimated as (10.8).

$$\text{yield} = -1091.3578^{**} + 278.8417^{***}W_1 + 17.0802^{***}N - 10.5563^{**}W_1^{1.5}$$
$$\qquad (403.44) \qquad (41.38) \qquad (2.992) \qquad (5.372)$$

$$\qquad - .6801^{***}N^{1.5} - .0970^{**}W_1N + 1042.3743^{**}pH_c - 2204.1327^{***}EC_c$$
$$\qquad\quad (.174) \qquad\quad (.046) \qquad\quad (322.80) \qquad\quad (116.14) \qquad (10.8)$$

$$\qquad + 2087.3238^{***}HC_c - 2.4966^{***}W_1PE2$$
$$\qquad\quad (185.44) \qquad\qquad (.515)$$

$$R^2 = .751 \qquad F = 86.29^{***}$$

Using (10.7) as the predicting equation, predicted yields and the 95 percent confidence limits for each predicted yield are given in Table 10.5. The average and range of observed yield for specified treatment combinations also are tabulated. The water treatment levels in Table 10.5 represent "middle range" levels.

For most experiments, average observed yields are reasonably close to their respective predicted yields. The most apparent exceptions are predicted yields for the Mesa experiment and the $(W = 32.1) \times (N = 0)$ treatment combination for the Safford experiment. Other than these exceptions, average observed yields tend to fall within the 95 percent confidence interval for predicted yields.

COTTON

The generalized production function for cotton (lint) based on experimental data from six sites is given in (10.9). Two weather variables were defined. $PE1$ represents the sum of daily pan-evaporation data from planting up to the first harvest. $PE1$ thus also reflects the length of the growing season. $PE2$ is defined as the sum of daily pan-evaporation data over a 20-day "critical period" during the flowering stage. The 20-day period was estimated to begin 105 days before the first harvest. This estimate represents a synthesis of comments by agronomists conducting the cotton experiment. Experimental data from the following six sites were used to derive (10.9): 1967 Shafter and 1969 West Side in California and 1971 Safford, 1971 Tempe, 1971 Yuma Mesa, and 1971 Yuma Valley in Arizona. Yield is expressed in pounds per acre of lint cotton.

$$\text{yield} = -626.1732^{***} + 96.8446^{***}W_1 + 1.685^{***}N - 30.9322^{***}W_1^{1.25}$$
$$\quad\quad (98.33) \quad\quad (11.72) \quad\quad (.642) \quad\quad (3.380)$$

$$\quad - .4780^{***}N^{1.25} + .0196^{***}W_1N - 70.4192^{***}pH - 8.9182^{*}EC$$
$$\quad\quad (.157) \quad\quad\quad (.005) \quad\quad\quad (25.21) \quad\quad\quad (5.232) \quad\quad \textbf{(10.9)}$$

$$\quad + 71.6196^{***}AWC + .1732^{***}W_1PE1$$
$$\quad\quad (5.261) \quad\quad\quad (.033)$$

$$R^2 = .802 \quad\quad F = 95.79^{***}$$

All estimated coefficients are statistically significant at the 10 percent or higher level. They also have the expected signs. When values for pH, EC, and AWC are coded, the 1.25 polynomial is reestimated as in (10.10). The properties of (10.9) and (10.10) are similar; the sign of the pH coefficient has changed. Using (10.9) as a predicting equation, predicted yields along with their 95 percent confidence limits are given for selected treatment combinations in Table 10.6. The water treatments represent middle range levels, that is, I_3. Predicted yields are relatively close to average observed yields.

$$\text{yield} = -800.2032^{***} + 102.1424^{***}W_1 + 1.8313^{***}N - 32.2791^{***}W_1^{1.25}$$
$$\quad\quad (103.42) \quad\quad (11.74) \quad\quad (.631) \quad\quad (3.339)$$

$$\quad - .5384^{***}N^{1.25} + .0215^{***}W_1N + 119.9998^{***}pH_c$$
$$\quad\quad (.157) \quad\quad\quad (.005) \quad\quad\quad (44.32) \quad\quad\quad \textbf{(10.10)}$$

$$\quad - 101.5129^{***}EC_c + 71.8776^{***}AWC + .1420^{***}W_1PE1$$
$$\quad\quad (36.25) \quad\quad\quad (5.202) \quad\quad\quad (.040)$$

$$R^2 = .805 \quad\quad F = 100.25^{***}$$

SUGAR BEETS (ROOTS)

In making preliminary analyses of generalized functions for roots, results were better when AWC rather than HC was included in the function. The generalized production function for roots based on experimental data from seven sites is given in (10.11). When agronomists conducting the experiments were asked to estimate the "critical growth period" for roots, there was not sufficient unanimity so as to define such a period. Therefore, $PE2$ represents the sum of daily pan-evaporation data from the date of planting up to three weeks prior to harvest. Thus $PE2$ also reflects the length of the growing season. Yield is expressed in tons per acre of roots adjusted to 15 percent sucrose. Experimental data from the following seven sites were used to derive (10.11); 1969 Ft. Collins and 1970 Walsh experiments in Colorado; the 1971 experiment at Plainview, Texas; and the 1971-72 (July harvest) Mesa, 1970 Safford, 1970-71 Yuma Mesa, and 1969-70 Yuma Valley experiments in Arizona.

$$\text{yield} = -34.3807^{***} - .6957^{***}W_1 + .0048N - .0184W_1^{1.5}$$
$$\quad\quad\;\; (3.933) \quad\;\; (.154) \quad\;\; (.009) \quad\;\; (.012)$$

$$- .0009^{**}N^{1.5} + .0006^{***}W_1N + 61.3137^{***}pH - 2.8004^{***}EC$$
$$\;\;(.0005) \quad\quad (.0001) \quad\quad (3.713) \quad\quad\quad (.354) \quad\quad \textbf{(10.11)}$$

$$+ 3.4588^{***}AWC - .2076^{***}PE2 + .0131^{***}W_1PE2$$
$$\quad (.272) \quad\quad\quad (.048) \quad\quad\quad (.002)$$

$$R^2 = .617 \quad F = 57.04^{***}$$

Snow and freezing weather around harvest adversely affected the sugar beet experiment at Ft. Collins. Consequently, these data were omitted, and the generalized function for roots at six sites was reformulated, as in (10.12).

$$\text{yield} = -114.8968^{***} + 1.1289^{***}W_1 + .0037N - .3104^{***}W_1^{1.25}$$
$$\quad\quad\;\; (6.720) \quad\quad\;\; (.294) \quad\quad (.017) \quad\quad (.092)$$

$$- .0181^{***}N^{1.25} + .0004^{**}W_1N + 58.1952^{***}pH - 3.4640^{***}EC$$
$$\;\;(.004) \quad\quad\quad (.0001) \quad\quad (4.797) \quad\quad\quad (.296) \quad\quad \textbf{(10.12)}$$

$$+ 5.5162^{***}AWC + .6340^{***}PE2 + .0955^{***}NpH$$
$$\quad (.315) \quad\quad\quad (.043) \quad\quad\quad (.028)$$

$$R^2 = .702 \quad F = 73.03^{***}$$

With the exception of N, all derived regression coefficients are statistically significant at the 1 percent level. One would hypothesize negative signs for the coefficients derived for pH and $PE2$. The positive sign for $PE2$ may represent a confounding of weather factors and length of growing season.

When the site variables assumed coded values, (10.12) was reestimated as (10.13).

$$\text{yield} = -120.7962^{***} + 3.0913^{***}W_1 + .1046^{***}N - .8422^{***}W_1^{1.25}$$
$$\quad\quad\;\; (8.086) \quad\quad\; (.336) \quad\quad\; (.027) \quad\quad\; (.105)$$

$$- .0268^{***}N^{1.25} + .0003^{*}W_1N + 48.9756^{***}pH_c - 8.6309^{***}EC_c$$
$$\;\;(.006) \quad\quad\quad (.0002) \quad\quad (4.627) \quad\quad\quad (.648) \quad\quad \textbf{(10.13)}$$

$$+ 29.6274^{***}AWC_c + 1.0277^{***}PE2 + .0509^{***}NpH_c$$
$$\quad (1.714) \quad\quad\quad\; (.065) \quad\quad\quad (.013)$$

$$R^2 = .612 \quad F = 48.63^{***}$$

Generalized function (10.12) thus is used for the predictions and comparisons in Table 10.7. Again as for the earlier generalized functions, the model seems generally useful for prediction at the spatially separated sites, which vary considerably in their associated soil and climatic environments.

PROMISE IN GENERALIZED FUNCTIONS

Research administrators, particularly those concerned with research for development of poor countries, have long expressed the need for investigating yield response results which have transferability among countries and world regions. Without such transferability, site specific research must be accomplished in each country or each region of it that differs with respect to soil and climatic variables. The Agency for International Development now places great emphasis on soil fertility, soil classification, and water management research that has promise of results that can be transferred among countries. The reason for this emphasis, of course, is either to (a) allow given research and aid funds to go further in knowledge generated and agricultural development attained or to (b) lessen the cost of obtaining a given set of knowledge and degree of development among countries.

The generalized functions developed in this chapter while far from "the final word" are very encouraging in these directions. In general, it was possible to estimate production functions that had reasonably accurate predicting power from production function experiments located at sites that had a great range in weather, soil, and spatial location. The study was, with respect to water response, a pioneering effort in this direction. Success can be attained, of course, only if the appropirate environmental variables (soil, climate, etc.) are specified and measured at each location prior to and in the duration of the experiments. Specification of more variables (others not listed were measured but did not prove useful in the estimation process) could have increased the predictive power of the generalized functions estimated in this chapter. Not only are those presented perhaps "primitive" in the number and kinds of environmental variables incorporated but also specification of the regression models might be highly upgraded in the presence of more data.

It must be recommended, however, that this was a small-scale project with a limited budget extending over four years. A budget in the millions of dollars extending over a decade or more would be more appropriate. (Some such projects of this type financed by AID do involve many millions of dollars and long time periods.) With a sufficiently large budget and time horizon, the environmental variables "accepted as given" for this study could be incorporated into a national or regional experimental design. This design could select locations so that there would be a great range of conditions (variables) representing various characteristics of soil, annual rainfall and its distribution, mean temperature, length of growing season, altitude, and others. The possibility of incorporating such variables into the experimental design, measurements, and regression models has been suggested by others for fertilizer.[4]

In the current study, the project was entirely voluntary and cooperative. The agronomists mentioned in an earlier chapter voluntarily conducted experiments at sites where they already had facilities under small grants from the Iowa State University Center for Agricultural and Rural Development, which held a contract with the Bureau of Reclamation for these and related purposes. The cooperators, who served so efficiently and ably, were all aware that the project had to be of short-run duration and only environmental variables that could be readily measured with facilities and equipment at hand were included. Even under these conditions and limitations the results were extremely promising. With appropriate budgets and planning horizons the task of generalized water response functions that can be used to predict yield outcomes over a wide range of soil and climatic conditions seems entirely feasible. Then, with this information in hand, irrigation projects can be better planned in terms of water allocation, area irrigated from given supplies, farm management plans, resource valuations and repayment periods, crop and livestock programs, market facilities, and community development.

DERIVED DEMAND FUNCTIONS FOR WATER

The traditional objectives or purposes for estimating water response or production functions are obvious and widely known: (a) to determine over what area a given water supply should be allocated, (b) from a given water supply, to determine how much should be allocated to soil areas that differ in characteristics and productivity, (c) to determine the optimal amount of water to be used per acre under either limited or unlimited supplies, (d) to determine how a given supply of water should be allocated among alternative crop and livestock activities if farm profit or regional income is to be maximized, (e) to determine the imputable values that are added to land and other resources as irrigation development takes place, (f) to aid in agricultural development and augment food supplies, and (g) other obvious reasons. Many of these applications from water response knowledge would be made through mathematical programming models or other appropriate methodologies rather than the classical economic principles and methods illustrated in Chapters 2 and 3. Several of these methodologies have been applied in this overall project but are not reported here because of space limitations and the emphasis per se on estimation of water response functions.[1]

Aside from these conventional applications of water response knowledge, water production functions also serve for other purposes. They are part of the data set involved in determining water demand and commodity supplies in irrigated agriculture. Farmers must use these quantities subjectively in determining their "real world" water demand and commodity supply functions. In a methodological sense, water demand and commodity supplies also can be derived on a normative basis through linear programming or other models.[2] To conserve space we, however, derive some water demand and commodity supply functions by classical economic methods in Chapters 11 and 12. These are static normative functions that illustrate further uses of water response functions. They are brief reviews illustrating methodology with the knowledge that their nature and elasticities will be functions of the types and models of water production functions upon which they are based.

NORMATIVE STATIC WATER DEMAND FUNCTIONS

In the preceding chapters, production functions were estimated for input-output relationships incorporating water and nitrogen. In this chapter, these relationships will be further analyzed through the estimation of short-run and longer-run demand functions for water. The same analyses could be completed for nitrogen. By specifying alternative price relationships and availabilities of the fixed inputs, the quantities of water that maximize the producer's objectives can be derived. Following determination of the optimum input use-level(s) and substitution of these into a production function, the corresponding output can be estimated. The quantities demanded and produced have implications for the size and structure of marketing mechanisms for both inputs and outputs. From another perspective, the projected availability and price of one input, for example, fertilizer, affect the demand for water and, in turn, the prices users are willing to pay. In a later chapter, water demand functions for individual crops are used in developing and quantifying a regional water allocation model. Given the specified demands for water, the impact of alternative water prices and distributions is examined.

The procedure for estimating the static demand for a single input was outlined in Chapter 2. Since the quantities of all other inputs used in the production process are considered fixed, the demand function is also termed a short-run function. The demand function is also normative in that the derived price-quantity relationships for the input being analyzed maximize the objective of the producer, for example, profit maximization. When one or more additional inputs are variable, the demand function will be referred to as a longer-run function.

Following a discussion of the derivation of short-run and longer-run demand functions, these functions are estimated for one set of experimental data selected from each of the corn, wheat, cotton, and sugar beet experiments analyzed in previous chapters. For mathematical convenience, the water demand functions are derived from polynomial functions of degree 2, that is, quadratic production functions. When other functional forms are being analyzed, the procedures are the same.

SHORT-RUN STATIC DEMAND FUNCTION

As noted in Chapter 2, a variable input can, in the absence of capital restraints and under perfect knowledge, be profitably used up to the point where the MVP of that input equals its price. This criterion assumes that the producer's goal is profit maximization and that the technical input-output relationships are estimable. A profit, π, function is in (11.1) where

$$\pi = P_y Y - P_1 X_1 - FC \tag{11.1}$$

$$Y = a + b_0 X_1 + b_1 X_1^2 \tag{11.2}$$

$$d\pi/dX_1 = (b_0 + 2b_1 X_1) = P_1 P_y^{-1} \tag{11.3}$$

$$X_1 = (P_1/P_y - b_0)/2b_1 \tag{11.4}$$

FC represents fixed costs and the production function is of the quadratic form, as in (11.2). In (11.3), the marginal value product (MVP) of X_1 equals its price and (11.4) is derived by rearranging the terms in (11.3). By assuming a value for P_y, the levels of X_1 corresponding to alternative values for P_1 can be used to plot a demand curve for X_1. As P_y, b_0, and (or) b_1 change, the shape and position of the demand curve for X_1 are altered.

LONGER-RUN STATIC DEMAND FUNCTION

Assume two inputs are variable so that another term, $-P_2 X_2$, is added to (11.1). Since $P_2 X_2$ was included in FC in the short-run situation, change FC to FC^*. A two-variable quadratic function is specified in (11.5). Since the coefficients for a one-variable and two-variable production function differ, the latter have asterisks. Recall that one of the conditions necessary for profit maximization is that the marginal rate of substitution between two inputs in (11.6) must equal the inverse of their price ratio.

$$Y = b_0^* + b_1^* X_1 + b_2^* X_2 + b_3^* X_1^2 + b_4^* X_2^2 + b_5^* X_1 X_2 \qquad (11.5)$$

$$\partial X_1/\partial X_2 = (b_2^* + 2b_4^* X_2 + b_5^* X_1)/(b_1^* + 2b_3^* X_1 + b_5^* X_2) = P_2/P_1 \qquad (11.6)$$

$$X_2 = [P_2 P_1^{-1}(b_1^* + 2b_3^* X_1) - b_2^* - b_5^* X_1]/[2b_4^* - P_2 P_1^{-1} b_5^*] \qquad (11.7)$$

The terms are rearranged so that X_2 is written in (11.7) as a function of levels of X_1, the technical coefficients and the price relationships. When (11.7) is satisfied, X_1 and X_2 are used in the correct combination. A demand function for X_1 can be developed by substituting (11.7) into (11.5), deriving dY/dX_1, equating this to P_1/P_y, and finally writing the resulting formulation in terms of X_1. The procedure will be developed in some detail for the corn experiment at Colby, Kansas.

Corn–Kansas, 1971

This experiment was extensively analyzed in Chapter 6. For convenience, the quadratic function in (6.3) is restated as (11.8). Using procedures outlined earlier, the short-run demand function for W is in (11.9), the quantified version of (11.4). Let $P_y = 0.04$ per pound and consider three levels of fertilization where $N = 90, 180,$ and 360 pounds per acre, respectively. The demand quantities of water under these alternative conditions are summarized in Table 11.1 and the relationships are plotted in Figure 11.1. Each demand curve is linear, and since each has the same slope $dW/dP_w = -1.245$, they are also parallel.

$$Y = -10586.0287 + 688.3354W + 36.4211N - 10.0386W^2$$

$$- .0772N^2 + .4133WN \qquad (11.8)$$

$$W = (688.3354 + .4133N - P_w P_y^{-1})/20.0772 \qquad (11.9)$$

The curves shift as different levels of nitrogen are incorporated or as different values are assumed for P_y. The price elasticity of demand, ϵ_w, P_w, for W with respect to P_w is an indicator of the responsiveness to demand quantity as the price paid by producers varies for water—in the static normative sense outlined earlier. This elasticity was defined in Chapter 2 and is restated in (11.10). Assume, for example, $N = 180$ and that P_w, the price of water, changes from 1.8 to 1.9, the arc or average elasticity throughout this price range is estimated by deriving dW from (11.9) and substituting in appropriate values for P_w and W, as in (11.11). A 1 percent increase in P_w, the price of

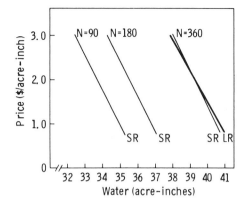

FIG. 11.1. Short-run (SR) and long-run (LR) normative
demand curves for water in the production of
corn for grain (Colby, Kansas, 1971).

water, is associated with only a 0.063 percent decrease in the demand quantity of W, water.

$$\epsilon_{w,P_w} = (dW/dP_w)(P_w/W) \qquad (11.10)$$

$$= [(35.6240 - 35.7485)/(1.9 - 1.8)]/[1.8/35.7485] = -.063 \quad (11.11)$$

When $N = 360$ and P_w increases from 1.8 to 1.9, ϵ equals -0.057. These low elasticities partly result from the fact that all inputs other than water are at fixed levels. In this experiment, the marginal product of water is so high that the price of water does not substantially affect the quantity of water demanded. The price elasticity with respect to P_Y, the price of corn, can also be readily derived. Assume $N = 360$, $P_w = 1.8$, and P_y increases by 15 percent from 0.04 to 0.046 per pound. The estimated price elasticity with respect to P_Y is 0.049. That is, a 1 percent decrease in P_y, corn price, is associated with only a 0.049 percent increase in the demand of water, W.

$$\epsilon_{W,P_y} = (dW/dP_y)(P_y/W) \qquad (11.12)$$

$$= [(39.7462 - 39.4539)/(.046 - .04)]/[.04/39.4539] = .049 \quad (11.13)$$

Using the steps earlier outlined for estimating a longer-run demand function for W, the MRS between W and N is in (11.14) and rewritten in terms of N in (11.15). Substituting (11.15) into (11.8), setting dY/dW equal to P_w/P_y, and rewriting the expression in terms of W, the longer-run demand function for W is in (11.16) where A represents $(.4133 + .1544 P_w P_n^{-1})$ and B is equivalent to $(20.0772 + .4133 P_w P_n^{-1})$.

$$\partial N/\partial W = (688.3354 - 20.0772W + .4133N)/(36.4211 - .1544N$$

$$+ .4133W) = P_w/P_n \qquad (11.14)$$

$$N = (-688.3354 + 20.0772W + 36.4211P_wP_n^{-1}$$

$$+ .4133P_wP_n^{-1}W)/(.4133 + .1544P_wP_n^{-1}) \tag{11.15}$$

$$W = \{688.3354 - P_wP_y^{-1} + [446.7447 + 30.1056P_wP_n^{-1}$$

$$+ (BA^{-1})(106.279 - 5.6234P_wP_n^{-1})]A^{-1}\}$$

$$/[20.0772 + .1544(BA^{-1})^2 - .8266BA^{-1}] \tag{11.16}$$

Assuming P_n is 0.1 per pound, the long-run (LR) demand curve in Figure 11.1 is estimated by assuming different levels for P_w and then deriving the corresponding values for W. The slope of the LR demand curve is less than that of the short-run curves, thereby suggesting that changes in P_w have a larger effect on quantities demanded, and the producer is more responsive to price changes. This is as expected. When a larger number of inputs becomes variable, the producer has more options in choosing combinations of inputs and, therefore, is more sensitive to changes in price of any one input. For example, the quantities of water and (or) nitrogen can be varied throughout a major portion of the production season. To test this hypothesis, compare the price elasticity of demand when P_w varies from 1.8 to 1.9 in the longer run to comparable estimates derived from the short-run formulations. The longer-run elasticity is estimated as -0.065 compared to -0.063 and -0.057 when $N = 180$ and 360, respectively, in the short run. While the longer-run elasticity is higher, the increase over the short-run estimations is not substantial.

Wheat—Yuma Valley, Arizona, 1971-72

The estimated quadratic function for this experiment is in (11.17) and the short-run water demand function is in (11.18). Let P_y equal 0.06 per pound and assume three levels of nitrogen. The profit-maximizing use-levels for water under these conditions are listed in Table 11.2 and plotted in Figure 11.2. The quantity of water demanded is not significantly affected by the price of water or the level of nitrogen. The formulation expressing the marginal product for water is in (11.19).

$$Y = -10414.9628 + 852.0111W + 11.6046N$$

$$- 12.9168W^2 - .0320N^2 + .0925WN \tag{11.17}$$

$$W = (852.0111 + .0925N - P_wP_y^{-1})/25.8336 \tag{11.18}$$

$$MP_w = 852.0111 - 25.8236W + .0925N \tag{11.19}$$

The coefficient for the N term is relatively small and (11.19) is not highly dependent on the level of fertilization. The MP of water is so high at all levels of N that changes in P_w are not of great importance.

Assuming $N = 150$, $P_y = 0.06$, and P_w increases from 1.8 to 1.9, the price elasticity of demand over this price range is estimated at -0.036. When N is increased to 250, other factors unchanged, the estimated price elasticity is

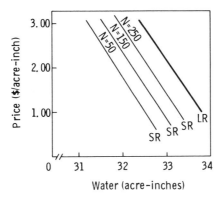

FIG. 11.2. Short-run (SR) and long-run (LR) normative
demand curves for water in the production of
wheat (Yuma Valley, Arizona, 1971–72).

essentially the same. With $N = 150, P_w = 1.8$, and P_y increasing from 0.06 to 0.075, the price elasticity of demand with respect to P_y is 0.029 over the specified price range.

In the longer-run context when the level of fertilization is permitted to vary, the price elasticity of demand over the water price range $P_w = 1.8$ to 1.9 is estimated as -0.034, which is nearly identical to the derived elasticities for the short-run situations.

Cotton—West Side, California, 1969

The estimated yield-water-nitrogen relationships for lint cotton are in (11.20) with the corresponding short-run water demand function in (11.21). Assuming the price for lint was 0.30 per pound, the quantities of water demanded under selected combinations of nitrogen applied and the price paid for irrigation water are listed in Table 11.3. The corresponding short-run demand curves are in Figure 11.3. When P_w is varied from 1.8 to 1.9 per acre-inch and N is at 60 and 180 pounds per acre, the estimated normative price elasticities of demand for water are -0.104 and -0.100, respectively. The price elasticity in the longer-run is only slightly higher at -0.107. The price elasticities with respect to changes in P_y are also low. When N is at $180, P_w$ is 1.8, and P_y increases from 0.3 to 0.4, the derived elasticity is 0.075. In the longer-run situation, this elasticity is slightly lower at 0.052. Since W and N are both variable in the longer run, a price elasticity higher than the short-run estimate would have been expected.

$$Y = -203.5364 + 62.8845W + 2.1906N - .9046W^2$$

$$- .0059N^2 + .0181WN \tag{11.20}$$

$$W = (62.8845 + .0181N - P_w P_y^{-1})/1.8092 \tag{11.21}$$

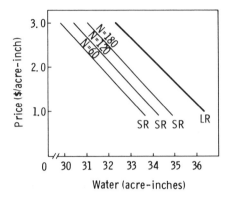

FIG. 11.3. Short-run (SR) and long-run (LR) normative
demand curves for water in production of cotton (lint) (West Side, California, 1969).

Sugar beets—Mesa, Arizona, 1971–72 (July)

The short-run water demand function (11.23) for this experiment was
derived from the quadratic formulation in (11.22). Using (11.23) as the
water demand function, the quantities of water demanded under alternative
levels of price and fertilization are depicted in Table 11.4 and Figure 11.4.
As with previous experiments, assume P_w increases from 1.8 to 1.9 dollars
per acre-inch. The derived price elasticities of demand when N is 180 and
360 pounds per acre are -0.075 and -0.067, respectively. With the longer-
run demand function in Figure 11.4, the elasticity over the same price range
for P_w is only -0.076. Finally, when P_y is increased from 18 to 20 dollars

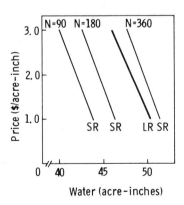

FIG. 11.4. Short-run (SR) and long-run (LR) normative
demand curves for water in the production of
sugar beets (Mesa, Arizona, 1971–72).

per ton, the elasticity is estimated as 0.067 when $N = 180$ and only 0.055 in the longer-run situation.

$$Y = -8.4863 + 1.2936W + .0600N - .0150W^2$$

$$- .00017N^2 + .00084WN \tag{11.22}$$

$$W = (1.2936 + .00084N - P_wP_y^{-1})/.03 \tag{11.23}$$

Qualifications in interpreting demand relationships

The quantities demanded and corresponding price elasticities are based upon normative demand functions. Prices and input-output relationships are implicitly assumed to be known with certainty. Since this rarely if ever occurs, quantities demanded are also affected by exogenous influences, for example, weather, and by producers' uncertainties regarding these external factors. No constraints have been imposed on availability of inputs. Also, profit maximization is the producers' assumed goal. When these assumptions are relaxed, the demand functions for water or any other input must be changed.

Most production units will have competing uses for water. Alternative crops can be grown. Some crops will be intermediate products to be used in livestock enterprises. When constraints on labor and capital availability are introduced, the relatively simple price-quantity demanded relationships developed for individual crops are less meaningful. They do, however, provide some insights. When demand functions are to be estimated in a more complex decision-making environment, linear programming and other estimating procedures are more applicable. This approach is discussed in Chapter 13.

Finally, the question of transferring relationships generated by field experiments to on-farm conditions must be acknowledged. Since experimental areas are relatively small, most field operations are conducted in a more timely manner than on-farm operations.

PRODUCT SUPPLY FUNCTIONS

In the previous chapter, demand functions were derived from the estimated production functions. This chapter is concerned with deriving static normative supply functions for crops, based on the estimated water response functions.

Considerable interest exists in estimating quantities produced and marketed, especially for major crops. Estimation of this supply component of conventional demand-supply analysis tends to be subject to more uncertainties than encountered in demand analysis. Producers make resource allocation decisions on the basis of expected price and growing conditions. These conditions, however, change throughout the production period. Even after making decisions as to what crops are to be grown and at what level of intensity, producers can vary quantities of water and fertilizer applied, rates of applying herbicides and pesticides, and for some crops, the harvest dates. The nature of supply response has important implications for forecasting quantities available for domestic consumption and for export, the relative and absolute levels of prices confronting processors and consumers, and the use-levels for marketing, transporting, and processing facilities. These conditions translate into consumer price indices, farm incomes, balance of payments position, etc.

In this chapter, normative short-run and longer-run product supply curves are estimated on the micro scale of this study. These are generated under rather special conditions. The curves are, for example, based on the assumption that input-output and price relationships are known with certainty. Also, the implicit assumption is made that each crop is exclusively grown and that no constraints on availability of inputs exist. Some of the qualifications necessary for interpreting the results are specified toward the end of the chapter.

SHORT-RUN PRODUCT SUPPLY FUNCTION

This analysis builds on that developed in Chapter 11. Recall that a single variable input can be profitably used up to the point where its marginal value productivity equals its cost. The corresponding demand function for this variable was developed by rearranging the terms in this profit-maximizing criterion. For reference, (11.4) is restated here as (12.1) where input is made a function of prices and other parameters. When (12.1) is substituted into (12.2), a short-run product supply function for Y is derived. By choosing a level for P_1, the price of input, and then varying the values for P_y, a supply curve for Y can be generated. The impact of changes in P_y occurs in (12.1).

$$X_1 = (P_1/P_y - b_0)/2b_1 \qquad (12.1)$$

$$Y = a + b_0 X_1 + b_1 X_1^2 \qquad (12.2)$$

That is, the intensity with which X_1 is applied is a function of several factors, including P_y, the price of the output. In subsequent sections, the level of fixed resources is varied by considering alternative levels of fertilization. A different product supply curve will be generated for each level of fertilization.

LONGER-RUN PRODUCT SUPPLY FUNCTION

Concepts developed in Chapter 11 will also be used in discussing derivation of a longer-run product supply function. With two variable inputs, as in (11.5), the MRS between the two is equated to their inverse price ratio, as in (11.14). At this point, the two inputs are used in the correct proportions. The profit-maximizing absolute level for water was specified in (11.16). The procedure is repeated so as to generate a longer-run demand function for the other input, N. The two equations, each of the form in (11.16), are then substituted into the production function (11.8), so that output is a function of the technical coefficients and of the relevant price ratios. Since the equation for a longer-run product supply function becomes very lengthy, it will not be stated here. While the procedure is somewhat cumbersome, it does incorporate the necessary conditions that both inputs are used in the correct proportions and absolute levels to maximize profits. By varying P_y while P_n and P_w are invariant, a longer-run product supply curve can be generated.

Corn—Kansas, 1971

The short-run water demand function, earlier developed as (11.9), is restated as (12.3). When (12.3) is substituted into the production function, (11.8), the resulting short-run product supply varies with the levels of N and $P_w P_y^{-1}$ where P_w is water price and P_y is corn price.

$$W_c = (688.3354 + .4133N - P_w P_y^{-1})/20.0772 \qquad (12.3)$$

By fixing P_w at, for example, 1.0, and assuming different levels of nitrogen, a family of product supply curves can be generated. When $N = 90$ pounds per acre, the supply function in (12.4) results. The quantities maximizing profits when $P_w = 1.0$ and P_y is varied are listed in Table 12.1. The procedure is repeated for other levels of fertilization.

$$Y_c = 18285.2094 - 36.1371 P_w P_y^{-1} - 10.0386(36.1373 - .0498 P_w P_y^{-1})^2$$

$$(12.4)$$

These relationships are also plotted in Figure 12.1. When N, nitrogen, is fixed at any level, the quantity produced and supplied is not responsive to the price received for the product. When N equals 90, for example, and P_y is tripled, the quantity produced and supplied only increases from 5113 to 5169 pounds per acre. But this is consistent with the demand analysis in the previous

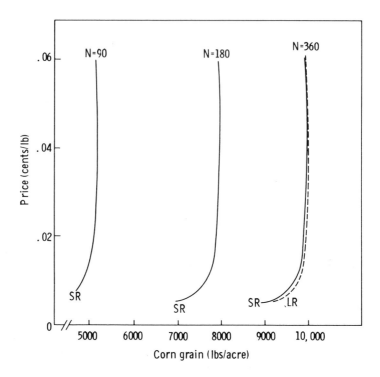

FIG. 12.1.   Short-run (SR) and long-run (LR) product supply curves for corn (grain) at specified prices
and levels of nitrogen (Colby, Kansas, 1971).

chapter. The demand function for water in (11.9) was not very responsive to
changes in P_w or P_y. That is, the short-run price elasticity of demand is low.
Given these conditions, the short-run product supply response will also be
low. The short-run product supply is, however, responsive to the level of
nitrogen. By doubling the rate of nitrogen from 90 to 180 pounds per acre,
the quantity produced increases by about 2780 pounds per acre.

$$\epsilon_{y,P_y} = [(5165.6953 - 5160.2016)/.01]/[.04/5160.2016] = .004 \qquad (12.5)$$

Price elasticities of supply are readily derivable. When N equals 90 and P_y
increases by 25 percent from 0.04 to 0.05, the short-run price elasticity of
supply in (12.5) is only 0.004 percent. That is, a 1 percent increase in P_y results in only a 0.004 increase in the quantity produced and supplied. When N
equals 360, other factors unchanged, the price elasticity of supply is only
0.002. These low elasticities are consistent with the shapes of the product
supply curves in Figure 12.1. Beyond $P_y = 0.02$, the curves are nearly per-
pendicular to the horizontal axis and, therefore, not sensitive to changes in
P_y. When P_y is again increased from 0.04 to 0.05, other factors unchanged,
the longer-run elasticity of supply is only 0.002.

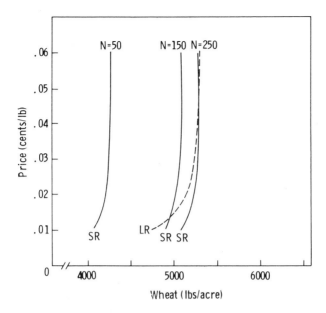

FIG. 12.2. Short-run (SR) and long-run (LR) product supply curves for wheat at specified prices and
levels of nitrogen (Yuma Valley, Arizona,
1971–72).

Wheat—Yuma Valley, Arizona, 1971–72

When the short-run demand function for water, (11.18), is substituted into the corresponding production function, (11.17), the quantity supplied–nitrogen-price relationships in Table 12.2 are generated. As with the Kansas corn experiment, the impact of P_y, wheat price, on quantity produced when all other inputs are fixed is not significant. The product supply curves in Figure 12.2 are steeply inclined and largely insensitive to levels of P_y.

As expected, the estimated price elasticities of supply are low. When N is at 50, P_w equals 1.0 and P_y is increased from 0.04 to 0.05 cent per pound, the price elasticity of supply is only 0.004. When $N = 250$, other factors unchanged, the price elasticity drops to 0.003. In the longer-run context, the price elasticity is estimated as 0.009 when P_y is increased from 0.04 to 0.05. When a lower portion of the long-run supply curve in Figure 12.2 is considered, for example, when P_y varies from 0.01 to 0.02, the price elasticity of supply is higher at around 0.082.

Cotton—West Side, California, 1969

Using procedures similar to those for estimating product supply curves for corn and wheat, quantities of cotton (lint) produced and supplied are sum-

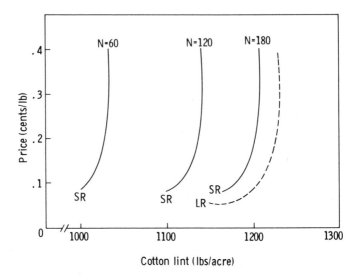

FIG. 12.3. Short-run (SR) and long-run (LR) product supply curves for cotton (lint equivalent) at
specified prices and levels of nitrogen (West
Side, California, 1969).

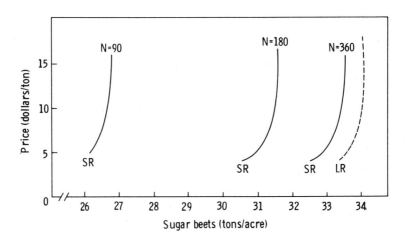

FIG. 12.4. Short-run (SR) and long-run (LR) product supply curves for sugar beets at specified prices and
levels of nitrogen (Mesa, Arizona, 1971–72,
July harvest).

marized in Table 12.3 and Figure 12.3. As with the other crops, the price elasticity of supply is low. When P_y is increased from 0.2 to 0.25 cent per pound and N is at 60 and 180, the estimated elasticities are 0.010 and 0.008, respectively. The long-run elasticity over this price range is only 0.006.

Sugar beets—Mesa, Arizona, 1971-72

Quantities supplied under alternative levels of price and nitrogen are in Table 12.4 and Figure 12.4. When P_y is increased by 25 percent from \$12 to \$15 per ton, water price or $P_w = 1.0$, and N is at 90 and 360, the price elasticities of sugar beet supplies on the micro basis analyzed are estimated at 0.006 and 0.005, respectively. Over the same price range, the long-run price elasticity of supply is only 0.002.

Qualifications in interpreting product supply curves

The product supply curves developed in this chapter are normative on an acre basis. They represent the optimum output levels under alternative price relationships for crops and water and levels of fixed inputs when the producer is attempting to maximize profits. Further, prices and input-output relationships are assumed to be known with certainty. Finally, each product has been considered in isolation; that is, the implicit assumption is made that the crop being analyzed is the only one being produced.

The features of these product supply curves will change as the factors affecting production are varied. Monthly constraints on water availability and family labor will likely affect the supply response to changes in P_y. When competing crops and livestock enterprises are introduced, changes in P_y for any product are expected to affect the relative profitability of each product. As the producer expects or experiences these changes, he may shift inputs to other uses. Even after the quantity produced is estimated, the product may be marketed over a period of time. Figures 12.1 through 12.4 are based on a static framework in that the quantity produced at a particular P_y is also marketed or supplied at that price. While the product supply analyses completed here are oversimplified, they do provide the concepts and insights that are incorporated into more complex decision-making environments.

PROGRAMMED WATER DEMANDS

Discussions in previous chapters focused on analyses of yield-water-nitrogen relationships and optimal water use within a microeconomic context. The quantities of water and (or) nitrogen that would be expected to maximize profits per acre, for example, were derived for a number of individual crops or production enterprises. Other focal points are of interest. The allocation of inputs, including water, within the farm or production unit represents a higher level of economic decision making. That is, the input allocation decisions within the production unit incorporate the relationships estimated for individual crops. Other levels of decision making could include input allocation within a watershed, state, or even national context. Analyses within these more macroeconomic contexts, however, require data on input-output relationships at the microeconomic levels.

Another segment of Contract 14–06–D–6192 with the Bureau of Reclamation involved a study of optimal water use within two watersheds in Emery County located in east central Utah. The study was done in a thesis by Jenson.[1] The watersheds are fed by two streams having high flows during the spring months but with receding flows throughout summer. Annual and seasonal flows also vary from year to year. As would be expected, seasonal flows do not coincide with water requirements, especially during midsummer and later summer. Water requirements listed in Table 13.1 equal or exceed average streamflows in both watersheds from July through October. This demonstrates one of the purposes for creating a reservoir system, namely, to capture and store those flows in excess of water requirements for later use when streamflows are insufficient to meet water needs. In this context, a recursive linear programming (LP) model was developed for estimating optimal allocation of water within the Emery County Project, given the physical features of the watersheds and the derived water requirements.

In building the regional model, some techniques described in earlier chapters are used. Cross-sectional data were used to estimate quadratic functions for four major crops grown on three soil classes within the project area. These production functions incorporate water as the only variable input applied to fixed levels of other resources. Such functions are of the form OA'' in Figure 3.1. A schedule of water demand or water requirements could be derived from each individual function (for example, barley grown on class 1 land). Each producer can grow crops in addition to barley and can include livestock enterprises within the constraints of labor, land, and capital availability.

Rather than develop water requirements for individual farms, a costly and time-consuming procedure, the concept of "representative farms" was used. Data on income from individual enterprises within the production unit were available. Based on the principal source of total value of all farm products sold during the year, farms were classified into one of the following groups: dairy (D), sheep (S), beef with yearlings fed on the farm (BF), and beef with calves sold in the fall (B). Farms were further classified into three sizes: small, medium, and large. With four types of farms, three sizes, and two watersheds, twenty-four representative farms were identified for the project area.

Yield-water coefficients were derived from the estimated production functions. Other input-output coefficients covering labor, machinery, and building requirements were developed from survey data collected within the project area. LP models were constructed for each of the twenty-four representative farms. Each model was then used to determine optimal resource allocation, including water, within the production unit. With the use of parametric programming, a water demand schedule was developed for each representative farm. These derived water requirements were later incorporated into the regional water allocation model.

NORMATIVE DEMAND CURVES FOR WATER

As noted earlier, parametric price programming was used to develop price-quantity combinations representing points on a demand curve for water. These demand curves are normative in the sense that they define the quantities of water that should be used at each price for water in order to maximize profits.

In estimating the demand curve in Figure 11.3, all inputs other than water were fixed. This situation could be readily incorporated into any LP analysis. Alternatively, the quantities of fertilizer bought and (or) labor hired can be permitted to change as the price of one input, for example, water, varies. Prices of fertilizer and hired labor, however, are invariant. This latter framework was used to estimate the normative demand curves for each of the twenty-four representative farms.

Water prices were varied from $0 to $12 per acre-foot. To demonstrate these adjustments to changing water prices, some of the results for medium-sized farm, beef-fed operations were abstracted. When the price of water falls from $12 to $0 per acre-foot, the optimum number of livestock units increase from 39 to 54. (See Table 13.2). The cropping pattern is relatively stable with a small increase in corn silage acreage at the expense of barley and rotation pasture acres. The marginal product of water in production of most crops is so high in this semiarid area that the price of water is relatively unimportant in affecting the optimal cropping plan. When the price of water drops to around $5.30 per acre-foot, irrigation of permanent pasture also becomes profitable. Increased forage from this land helps sustain the higher number of livestock.

When water is priced at $12 per acre-foot, about 176 acre-feet are demanded or required for profit maximization. When water is free, about 453 acre-feet can be profitably used. These two price-quantity combinations represent two points on the "stepped" normative demand curve for water depicted in Figure 13.1. These and the other points on the "stepped" demand curve were derived through parametric price programming.

The other points in Figure 13.1, that is, those not connected by line segments, represent the anticipated distribution of water according to contractual agreement and under the Bureau of Reclamation's rule that the available water should be allocated in proportion to the number of irrigable acres. When water is priced above $5.30 per acre-foot, the production unit in Figure 13.1 would profitably use less water than the quantities distributed. When the price is $5.30 or less making irrigation of permanent pasture profitable, the quantity demanded at all prices exceeds the quantity that would be distributed under the contractual agreement. With water priced at $5.30, the difference between the quantity demanded and distributed is around 134 acre-feet. The magnitude and sign of this difference vary among representative farms. This suggests that producers could improve their individual and collective positions through a redistribution of available water.

The derived demand for each representative farm was expanded by a weight representing the corresponding incidence of that farm in the project area. These quantities demanded were then aggregated for the Cottonwood

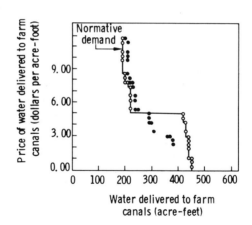

FIG. 13.1. Derived normative demand curve for water on medium-sized beef operation in the Cottonwood watershed compared to anticipated water distribution under existing distribution procedures. (From Jenson, *Programming Models,* Fig. 15, 1971).

and Huntington watersheds. The magnitudes of the aggregated quantities demanded are reflected in the following price-quantity demanded relationships:[2]

Price (dollars/acre-foot)	Cottonwood (acre-feet)	Huntington (acre-feet)
12.00	10002	25684
6.02	11826	31678
2.90	18794	44535
0.32	22441	50205

The relatively large increase at points in the $3 to $6 range results from the profitability of irrigating permanent pasture. As noted earlier, when the price of water is $5.30 or less, irrigation of permanent pasture is profitable on a medium-sized beef-feeding (BF) farm in the Cottonwood watershed. In addition to water for agricultural purposes, water for domestic, municipal, and industrial uses was included in the analyses. The requirements for these latter uses were fixed, that is, independent of the variable pricing for deriving the demand among representative farms. Water demanded for all uses was also specified in terms of monthly requirements.

WATER SUPPLIES

The other major component in constructing the regional water allocation model is a specification of water availability and water transfer systems. Water is supplied by two creeks and storable in two major and several smaller reservoirs. Average flows into the two watersheds were included in Table 13.1. The regional model also includes specification of reservoir capacities; locations, capacities, and efficiencies of the water conveyance systems; and discharge mechanisms. Allowances are also made for maintenance of non-active reservoir supplies and for water losses through evaporation and drainage.

As apparent, the reservoir system permits storage of excess flows for use later in the production year and (or) subsequent years. Given the estimated monthly and annual water demands, the objective is to determine the optimal set of strategies for storing and discharging available water so as to maximize, for example, total farm income over some planning period.

BASIC REGIONAL WATER ALLOCATION MODEL

Recall that the demand for water for agricultural uses was derived for four consuming areas, three in the Huntington and one in the Cottonwood watersheds. Also, a fixed level of water was required for industrial, municipal, and domestic uses. The model was constructed to maximize total "consumer surplus" subject to water availability and the physical feature of the water

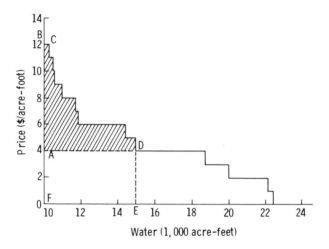

FIG. 13.2. **Derived demand for water in the Cottonwood area.** (From **Jenson,** *Programming Models,* 1971)

storage and conveyance systems. The term "consumer surplus" represents the difference between the amount the user would be willing to pay and what he actually pays. Referring to Figure 13.2, when the price of water is $4.0 and about 15,000 acre-feet are demanded, the collective user pays an amount equal to *ADEF*. Since the user would be willing to pay *FBCDE*, the area representing the consumer surplus is *ABCD*.

Three similar demand schedules but of different shape and position could be plotted for the three consuming regions in the Huntington watersheds. Estimates of respective consumer surpluses corresponding to alternative prices could also be readily derived.

The thrust of the model can be demonstrated by considering two time periods. This construct is generalizable to a number of years. The objective is to maximize the consumer surplus within the four consuming regions over the two-period planning horizon. The usual quantity demanded is a function of price relationship, $q = f(p)$, and is inverted so that $p = \phi(q)$ or $p = \phi(w)$.

$$\text{Max} \sum_{t=1}^{2} \sum_{j=1}^{4} \sum_{w_{ijt}=0}^{w_{jt}^{o}} \phi w_{ijt} dw_{ijt} - p^{o} w_{jt}^{o} \tag{13.1}$$

subject to

$$\sum_{t=1}^{2} w_{jt} \leqslant \overline{W} \quad j = (1, \ldots, 4)$$

and

$$\overline{W} = W_t + W^*{}_{t-1}$$

$$w_{jt} \geqq 0$$

where w_{jt}^o = quantity of water demanded in the jth consuming area in period t
        when price equals p^o;

   $\overline{W}$ = maximum quantity of water available over the two-period plan-
        ning horizon net of nonagricultural water requirements;

   W_t = streamflows in period t net of nonagricultural water require-
        ments; and

W^*_{t-1} = water stored in reservoirs at the end of the $t-1$ time period.

The maximization of (13.1) is, of course, also subject to constraints on the

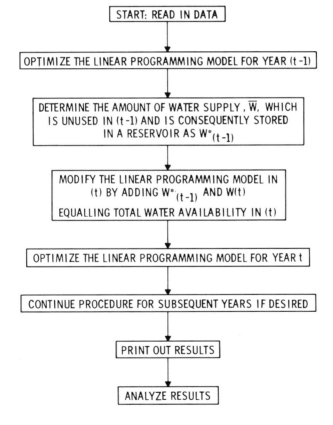

FIG. 13.3.   Schematic diagram of a simple recursive linear
        programming model. (From Jenson, *Program-
        ming Models,* 1971)

water conveyance systems. In terms of Figure 13.2, the objective is to allocate available water over space and through time so as to maximize the total shaded portions for the four demand functions over a two-period planning horizon. The sequence of analytical steps is summarized in Figure 13.3.

RESULTS FROM MODEL

The model was simulated for 1920–59, the period for which "historical" water flow data were available. The demand curves for each of the four consuming areas are assumed to be applicable to each year over the 1920–59 period. Two sets of results were generated: one for water allocation without implementation of the Emery County Project and the other with the project. In the "preproject" situation, the two watersheds were independent; that is, there were no canals transferring water between the two watershed areas. Also, the dam and reservoir system was not in effect. With the "project" setting, water storage for interspatial and intertemporal use is available as well as water conveyance systems between the Cottonwood and Huntington watersheds.

The major results of applying the regional water allocation model under the "preproject" and "project" conditions are summarized in Table 13.3. Estimated water shortages were more prevalent in the Huntington than the Cottonwood area. But this is consistent with the data in Table 13.1. In the Huntington watershed, shortages would have been expected in 31 of 40 years without the project but in only 3 years with the project. Even though several of the shortages under "preproject" conditions are relatively small, the development of a reservoir system and canals for transferring water from the Cottonwood watershed would have permitted substantial reductions in annual and total water shortages over the 1920–59 period. Shortages in the Cottonwood watershed were generated for 12 of the 40 years under preproject conditions and only 3 years with the project.

The estimates of the marginal value product, MVP, of water represent the addition to total consumer surplus if an additional acre-foot of water were available. In 1956, for example, the value of an additional acre-foot of water under preproject conditions in the Huntington area is estimated to be $4.96 compared to only $0.88 after implementation of the project. The comparable estimates for the Cottonwood area are $1.56 and $0.79, respectively. Recall that in the "preproject" situation, no water transfers occurred between the two watersheds. Consequently, the MVP of water in Huntington, the watershed with more frequent and larger shortages, is considerably higher than that in the Cottonwood area for comparable years. The exception is 1924. When interwatershed transfers of water are possible, the MVP between the two areas should tend to be equalized. That is, water is transferred and used where it is most remunerative. Following creation of the project, the MVPs between the two areas are more comparable. They are not, however, identical because movement of water is not fully free but constrained by the water transfer opportunities implicit in the reservoir and canal system.

Jenson also employed variable resource programming in estimating cost-benefit analyses of alternative modifications to the water reservoir and transfer system. Several levels of reservoir capacity, for example, were considered. None of these was determined to be economically remunerative even at interest rates as low as 4 percent.

CHAPTER 1

1. See Uri Horowitz, A Dynamic Model Integrating Demand and Supply Relationships for Agricultural Water, Applied to Determining Optimal Intertemporal Allocation of Water in a Regional Water Project. Ph.D. dissertation, Iowa State Univ., Ames, 1974.

2. See Earl A. Jenson, Programming Models of Irrigation Development. Ph.D. dissertation, Iowa State Univ., Ames, 1971.

3. Chris Beringer, An Economic Model for Determining the Production Function for Water in Agriculture. Calif. Agr. Exp. Sta. Report 240, Berkeley, 1961.

4. C. H. Wadleigh, The Integrated Moisture Stress upon a Root System in a Large Container of Saline Soil. Soil Sci. 61:225–38, 1946.

5. S. A. Taylor, Estimating the Integrated Soil Moisture Tension in the Root Zone of Growing Crops. Soil Sci. 73:331–40, 1952.

6. See Metin Caglar, Optimization of Intraseasonal Water Allocation. Ph.D. dissertation, Iowa State Univ., Ames, 1974. Also see Horowitz, A Dynamic Model.

CHAPTER 2

1. For a discussion of the stages of production, see Earl O. Heady, Economics of Agricultural Production and Resource Use (Englewood Cliffs, N.J.: Prentice-Hall, 1952), Chapter 4.

2. Price supports and controls reduce price uncertainty. For a discussion of price expectation models, see Earl O. Heady, Economics of Agricultural Production and Resource Use (Englewood Cliffs, N.J.: Prentice-Hall, 1952), pp. 475–96.

CHAPTER 3

1. P. J. Kramer, Plant and Soil Water Relationships (New York: McGraw-Hill, 1949), p. 37.

2. D. W. Thorne and H. B. Peterson, Irrigated Soils (New York: Blakistone, 1954), cited in Chris Beringer, Some Conceptual Problems in Determining the Production Function of Water. Proc. Western Farm Econ. Assoc., July 14–17, 1959, p. 60.

3. Beringer, Some Conceptual Problems, pp. 60–61.

4. W. L. Parks, Methodological Problems in Agronomic Research Involving Fertilizer and Moisture Variables, in E. L. Baum, E. O. Heady, and J. Blackmore, Methodological Procedures in the Economic Analysis of Fertilizer Use Data (Ames: The Iowa State College Press, 1956), pp. 113–33.

5. Chris Beringer, An Economic Model for Determining the Production Function for Water in Agriculture. Calif. Agr. Exp. Sta. Report 240, Berkeley, 1961.

6. C. V. Moore, A General Analytical Framework for Estimating the Production Function for Crops Using Irrigation Water. J. Farm Econ. 43 (1961): 876–88.

7. Moore, A General Analytical Framework, p. 877.

8. R. H. Shaw and R. E. Felch, Climatology of a Moisture-Stress Index for Iowa and Its Relationship to Corn Yields. Iowa State J. Sci. 46 (1972): 357–68; O. T. Denmead and R. H. Shaw, The Effect of Soil Moisture Stress at Different Stages of Growth on the Development and Yield of Corn. Agron. J. 52 (1960): 272–74.

9. Beringer, An Economic Model.

10. Ibid.

11. Moore, A General Analytical Framework.

12. C. W. Cobb and P. H. Douglas, A Theory of Production. Am. Econ. Rev. 18(1928): 139–65.

13. C. D. Hoover, Applicability of the Mitscherlich Method to Determination of Phosphorus Deficiencies of Iowa Soils. Ph.D. dissertation, Iowa State College, Ames, 1939.

14. G. E. Briggs, Plant Yield and Intensity of External Factors— Mitscherlich's Wirkungsgesetz. Ann. Bot. 39(1925): 475–502; Hoover, Applicability of the Mitscherlich Method.

15. Briggs, Plant Yield.

16. B. Baule, Zu Mitscherlich's Gesetz der Physiologischen Beziehungen. Landw. Jahrb. 51 (1918): 363–85.

17. Hoover, Applicability of the Mitscherlich Method.

18. B. H. Balmukand, Studies in Crop Variation. V. The Relation Between Yield and Soil Nutrients. J. Agr. Sci. 18(1928): 602–27.

19. Briggs, Plant Yield.

20. W. J. Spillman, Application of the Law of Diminishing Returns to Some Fertilizer and Feed Data. J. Farm Econ. 5:36–52.

21. See, for example, E. O. Heady and J. L. Dillon, Agricultural Production Functions (Ames: Iowa State Univ. Press, 1961).

22. R. G. D. Allen, Mathematical Analysis for Economists (London: Macmillan, 1962), p. 447.

23. Ibid., p. 453.

24. Shaw and Felch, Climatology of a Moisture-Stress Index; W. C. Corsi and R. H. Shaw, Evaluation of Stress Indices for Corn in Iowa. Iowa State J. Sci. 46(1971):79–85; O. T. Denmead and R. H. Shaw, Availability of Soil Water to Plants as Affected by Soil Moisture and Meteorological Conditions. Agron. J. 54(1962):385–90.

25. J. C. Flinn, Allocation of Water Resources: A Derived Demand Function for Irrigation Water, and Efficiency of Allocation. Farm Mgt. Bull. 1, Armidale, Australia, U. of New England, 1968.

26. N. J. Dudley, A Simulation and Dynamic Programming Approach to Irrigation Decision Making in a Variable Environment. Agr. Econ. Bus. Mgt. Bull. 9, Armidale, Australia, U. of New England, 1969.

27. See Larry A. Nelson and R. L. Anderson, A Family of Models and Experimental Designs for Evaluating Response to Fertilizer Nutrients When the True Model Is Not Known. Paper Presented at Soil Fertility and Classification Conference, Univ. of Hawaii, May 19, 1974; R. L. Anderson and Larry A. Nelson, Some Problems in Estimation of Single Nutrient Response Functions. Paper 4B, Department of Statistics, Univ. of Kentucky, Lexington.

28. Wayne A. Fuller, Grafted Polynomials as Approximating Functions. Australian J. Agr. Econ. 13 (1969):35–46.

29. See Metin Caglar, Optimization of Intraseasonal Water Allocation. Ph.D. dissertation, Iowa State Univ., Ames, 1974; Earl A. Jensen, Programming Models of Irrigation Development. Ph.D. dissertation, Iowa State Univ., Ames, 1971.

CHAPTER 4

1. O. Kempthorne, The Design and Analysis of Experiments (New York: John Wiley, 1952).

2. W. G. Cochran and G. M. Cox, Experimental Designs (New York: John Wiley, 1950); D. R. Cox, Planning of Experiments (New York: John Wiley, 1958); W. T. Federer, Experimental Design (New York: Macmillian, 1955); R. A. Fisher, The Design of Experiments (London: Oliver and Boyd, 1951); Kempthorne, The Design and Analysis; B. Ostle, Statistics in Research (Ames: Iowa State Univ. Press, 1963).

3. Ostle, Statistics in Research, p. 247.

4. Cox, Planning of Experiments, p. 95.

5. Federer, Experimental Design, Chapter IV.

6. Typically, however, estimated response functions for fertilizer are in Stage 2 where the first derivative or marginal product is declining. Often or typically the same is true for water.

7. P. Rao and R. L. Miller, Applied Econometrics (Belmont, Calif.: Wadsworth, 1971).

8. N. R. Draper and H. Smith, Applied Regression Analysis (New York: John Wiley, 1966).

9. J. Johnston, Econometric Methods (New York: McGraw-Hill, 1963), p. 207.

10. H. O. Hartley, The Modified Gauss-Newton Method for Fitting of Nonlinear Regression Functions by Least Squares. Technometrics 3(1961): 269–80.

11. O. L. Davies, ed., Design and Analysis of Industrial Experiments, 2nd ed. (New York: Hafner, 1956).

12. D. W. Marquardt, An Algorithm for Least Squares Estimation of Nonlinear Parameters. J. Soc. Ind. Appl. Math. 2(1963):431–41.

CHAPTER 5

1. Chris Beringer, Some Conceptual Problems in Determining the Production Function of Water. Proc. Western Farm Econ. Assoc., July 14–17, 1959, p. 58.

2. Roger W. Hexem, Earl O. Heady, and Metin Caglar. A Compendium of Experimental Data for Corn, Wheat, Cotton, and Sugar Beets Grown at Selected Sites in the Western United States and Alternative Functions Fitted to These Data. Special Research Report, Center for Agricultural and Rural Development, Iowa State Univ., Ames, June 1974.

CHAPTER 6

1. The concept and use of dummy variables were discussed in Chapter 4.

CHAPTER 8

1. D. W. Grimes and R. M. Hagan, Crop Response Functions to Irrigation, Fertilizer, and Other Production Factors . . . Cotton, mimeographed. Dept. of Water Science and Engineering, Univ. of California at Davis, 1970, p. 25.

CHAPTER 9

1. H. D. Vinod, Econometrics of Joint Production. Econometrica 36(1968):322–35.

2. H. Hotelling, Relations between Two Sets of Variates. Biometrika 28(1936):321–77.

3. M. Kaminsky, The Structure of Production of Multiple-Output Dairy Farms in the "Centro Santafecino" Region of Argentina: A Multivariate Analysis. Ph.D. dissertation, Univ. of Wisconsin, June 1971.

4. T. W. Anderson, An Introduction to Multivariate Statistical Analysis (New York: John Wiley, 1958).

5. James S. Press, Applied Multivariate Analysis (New York: Holt, Rinehart and Winston, 1972).

6. Y. Mundlak, Transcendental Multiproduct Production Functions. Int. Econ. Rev. 5(1964):273.

7. L. R. Klein, A Textbook of Econometrics (New York: Harper and Row, 1953).

8. Alan A. Powell and F. H. Gruen, The Constant Elasticity of Transformation Production Frontier and Linear Supply System. Int. Econ. Rev. 9(1968):322.

9. R. G. D. Allen, Mathematical Analysis for Economists (Macmillan, London, 1964).

10. Kaminsky, The Structure of Production.

11. Ibid.

CHAPTER 10

1. Soil Conservation Services, U.S. Department of Agriculture, Guide for Interpreting Engineering Uses of Soil (Washington, D.C.: U.S. Government Printing Office, 1971).

2. F. B. Cady and W. A. Fuller, The Statistics-Computer Interface in Agronomic Research. Agron. J. 62(Sept.–Oct.):599–604. 1970.

3. Regis Voss and John Pesek, Geometrical Determination of Uncontrolled-Controlled Factor Relationships Affecting Crop Yield. Agron. J. 57(1965):460–63.

4. For example, see the following: R. E. Voss, J. J. Hanway, and W. A. Fuller, Influence of Soil Management and Climatic Factors on the Yield Response by Corn to N, P and K Fertilizer. Agron. J. 62(1970):736–40; R. J. Laird and F. B. Cady, Combined Analysis of Yield Data from Fertilizer Experiments. Agron. J. 61:829–34; F. B. Cady and D. M. Allen, Combining Experiments to Predict Future Yield Data. Agron. J. 64:211–14.

CHAPTER 11

1. See Uri Horowitz, A Dynamic Model Integrating Demand and Supply Relationships for Agricultural Water, Applied to Determining Optimal Intertemporal Allocation of Water in a Regional Water Project. Ph.D. dissertation, Iowa State Univ., Ames, 1974; Earl A. Jenson, Programming Models of Irrigation Development. Ph.D. dissertation, Iowa State Univ., Ames, 1971; and Metin Caglar, Optimization of Intraseasonal Water Allocation. Ph.D. dissertation, Iowa State Univ., Ames, 1974.

2. For example, see Jenson, Programming Models.

CHAPTER 13

1. Earl A. Jenson, Programming Models of Irrigation Development. Ph.D. dissertation, Iowa State Univ., Ames, 1971.

2. Ibid., Table 125.

INDEX

Canonical correlation, 162–67
Climatic factors. *See* Weather factors
Coefficient of determination, 63
Corn experiments
Arizona
Mesa, 1970
corn grain
square root function, 96
three-halves polynomial, 96
corn silage, three-halves polynomial, 97
Mesa, 1971
corn grain, three-halves polynomial, 94–95
corn silage, square root function, 95
Safford, 1972, quadratic function, 100
Yuma Mesa, 1970
corn grain, quadratic function, 97–98
corn silage, quadratic function, 98
Yuma Valley, 1970
corn grain
square root function, 99
three-halves polynomial, 99–100
corn silage, three-halves polynomial, 100
California
Davis, 1969
quadratic function, 92
square root function, 93
Davis, 1970, square root function, 91–92
Colorado, Ft. Collins, 1968
corn grain
Mitscherlich model, 90
three-halves polynomial, 88–89
corn silage, quadratic function, 90–91
Kansas
Colby, 1971
Mitscherlich model, 80
quadratic function, 78–80
isoclines, 84–86
isoquants, 84–86
marginal product, 82–84
marginal rate of substitution, 85–86
production surface, 80–81
Colby, 1970, three-halves polynomial, 87–88
Texas
Plainview, 1969, quadratic function, 103–4
Plainview, 1970, three-halves polynomial, 102–3
Plainview, 1971, three-halves polynomial, 101–2
Plainview (Lake Site), 1971, square root function, 104–5

Cotton experiments
Arizona
Safford, 1971, quadratic function, 133
Tempe, 1971, three-halves polynomial, 133–34
Yuma Mesa, 1971, three-halves polynomial, 134
Yuma Valley, 1971, square root function, 135
California
Shafter, 1967–69, pan evaporation, 123
Shafter, 1968, quadratic function, 122
Shafter, 1969, quadratic function, 121
average product, 121
Shafter, 1967, square root function, 122–23
Shafter, 1967–69, three-halves polynomial, 124–25
West Side, 1967–68, pan evaporation, 132
West Side, 1967, square root function, 131–32
West Side, 1968, square root function, 131
West Side, 1969, square root function, 125
isoquant, 128–29
marginal product, 125–29
marginal rate of substitution, 128, 130
production surface, 125–26

Dynamic functions, 27–28
Dynamic programming, 28

Elasticity of production, 43
Experimental design, 74–75
completely randomized, 56–57
experimental error, 52–53
blocking, 52–53
factorials, 54–55
financial and time constraints, 55
objectives, 48–52
plotting, 59–60
quality of management inputs, 53–54
randomization, 54
randomized complete block design, 57
randomized incomplete block design, 58
replication, 54

Factors
effect, 37
growth, 36–37
injury, 38

Irrigation scheduling
evaporation, 34

Irrigation scheduling *(continued)*
  plant transpiration, 34
  tensiometer, 34

Joint products, 162

Linear programming, 28
LOFF, defined, 64–65

Multicollinearity, 68

Point estimates, 10–11
Production functions
  constraints in algebraic forms, 43–45
  economic applications, 9–28
  estimation of, 60–69
  features, 29–45
  field capacity, 30
    gravitational water, 31
    irrigation scheduling, 34–35
    permanent wilting point, 30
    plant growth, 32–34
      equal availability theory, 32
    plant-water relationship, 29–35
    soil moisture tension, 30
    water-holding capacity, 29–32
  generalized
    aggregation bias, 175
    corn, grain
      1.25 polynomial, 176–78
      three-halves polynomial, 177, 174
    corn, silage, square root function, 180
    cotton, 1.25 polynomial, 182
    ordinary least-square method, 176
    sugar beets, roots
      1.25 polynomial, 183
      three-halves polynomial, 183
    wheat, three-halves polynomial, 181
  joint, 162-64
    Cobb-Douglas function, 164–65
    constant elasticity of substitution and
      transformation (CEST) function,
      165–68
    marginal product, 170–71
    marginal rate of substitution, 171
    marginal rate of transformation, 171
    corn experiments, 172–73
    quadratic function, 165–67, 168
      marginal product, 169
      marginal rate of substitution, 169–70
      marginal rate of transformation,
        169–70
    transcendental function, 164, 167–68
  multivariable inputs, 19–27
    economic optimum input use levels,
      23–25
    expansion path, 23–40
    isoclines, 22, 36, 38, 40, 42, 44
    marginal rate of substitution, 21, 22,
      36, 38, 40, 42
    price elasticity of supply, 26, 27
    product isoquant, 19, 36, 38, 40
    product supply function, 25–28

profit function, 23, 24, 26
ridge lines, 22, 40
one variable input, 11–19
  average product, 12, 44
  economic optimum input use level,
    13–14
  marginal product, 12, 36, 40, 42, 44
  ordinary least-square method, 62
  price elasticity of demand, 16
  production possibility curve, 17
  rate of product transformation, 18
  static demand, 15
  role in management, 3–8
    decision on water use, 3
    environmental condition, 6
  specification, 35–43
    Cobb-Douglas function, 35–36, 43
    Mitscherlich-Spillman function, 36–39,
      44
    polynomial functions, 39–40
      Taylor's series expansion, 39–40
    quadratic function, 35, 39–40, 44
    square root function, 42, 44
    three-halves polynomial function, 42
  specification bias, 61
  transferability of data, 47

Response functions. *See* Production
  functions

Simulation, 28, 43
Soil characteristics, 173–74
Sugar beet experiments
  Arizona
    Mesa
      July harvest, 1969–70
        root yield function, 148
        top yield function, 148–49
      July harvest, 1969–72
        root yield function, 152
        top yield function, 152–53
      July harvest, 1970–71
        root yield function, 146
        top yield function, 146–47
      July harvest, 1970–72
        root yield function, 153
        top yield function, 153
      July harvest, 1971–72
        root yield function, 138–43
          isoclines, 142–43
          isoquants, 142
          marginal product, 141–43
          marginal rate of substitution,
            142–43
          production surface, 139
        top yield function, 143–44
      May harvest, 1969–70
        root yield function, 147
        top yield function, 147–48
      May harvest, 1969–72
        pan evaporation, 149
        root yield function, 150
        top yield function, 151

May harvest, 1970–71
  root yield function, 145
  top yield function, 145
May harvest, 1970–72
  root yield function, 152
  top yield function, 152
May harvest, 1971–72
  moisture tension, 136
  root yield function, 137
  top yield function, 137–38
Safford, 1970, root yield function, 159
Yuma Mesa, 1969–70
  root yield function, 155–56
  top yield function, 156
Yuma Mesa, 1970–71
  root yield function, 154
  top yield function, 154
Yuma Valley, 1969–70
  root yield function, 158
  top yield function, 158–59
Yuma Valley, 1970–71
  root yield function, 157
  top yield function, 157–58
Colorado
Ft. Collins, 1969, root yield function, 160
Walsh, 1970, root yield function, 160–61
Texas
Plainview, 1971, root yield function, 161–62
Supply function
corn
  long-run, 196–97
  short-run, 195–96
cotton, 197–98
long-run, 195
short-run, 194–95
sugar beets, 199
wheat, 197

Treatments
  fertilizer, 75
  irrigation, 75
t-test, 67

Water
  basic regional allocation model, 203–7

derived demand function
corn
  long-run, 189–90
  short-run, 188–89, 195
cotton
  long-run, 191–92
  short-run, 191–92
long-run static, 188
short-run static, 187
sugar beets
  long-run, 192–93
  short-run, 192–93
wheat
  long-run, 191
  short-run, 190–91
dynamics of response, 7
marginal productivity, 5–6
normative demand, 201–3
resource planning, 4
response function, 6
supplies, 203
Weather factors, 174–75
Wheat experiments
Arizona
  Mesa, 1970–71, three-halves polynomial, 117
  Mesa, 1971–72
    quadratic function, 116
    square root function, 116
  Safford, 1970–71, quadratic function, 118
  Yuma Mesa, 1970–71
    quadratic function, 115
    square root, 114–15
  Yuma Mesa, 1971–72, quadratic function, 113–14
  Yuma Valley, 1970–71, three-halves polynomial, 112–13
  Yuma Valley, 1971–72
    Mitscherlich function, 107
    quadratic function, 106–7
    isoquant, 110–11
    marginal product, 109–10
    marginal rate of substitution, 111–12
    production surface, 107–8
Colorado
  Walsh, 1970–71, quadratic function, 118–19